I0469245

I

L

O

V

E   B

E   V

Did an encounter with a UFO forty years ago alter John's brain in such a way that he can now answer the questions to so many of the riddles that have eluded us? Among other things, Mr. Atkins gives us:

-a more accurate depiction of the atom;
-a theory of planetary pairing;
-the formulas for Unified Field;
-an explanation for blackholes;
-the speed of light is near infinite;
-a formula for gravity;
-an exact explanation of gravity;
-an amazing explanation for the solar system!

That any man could come up with even one formula or explanation would be astonishing. That one man came up with a precise explanation for the whole solar system, gravity, and the atom, is absolutely incredible. You just cannot read this book and walk away without knowing that you have just witnessed something spectacular. This ranks as one of the most remarkable achievements in the history of man.

WOW!!!

No Force of Attraction

by John Hildreth Atkins

<u>Dedication</u>

I dedicate this book to all the free-thinkers in the world who challenge that which so-called learned men would have them believe. I do this with all the fervor and ardor that only a repressed man can appreciate after a prolonged period of censorship.

Table of Contents

Part Two
(my earlier thoughts)

INTRODUCTION

In my autobiography, *Mi Vida Loca-Walk a Mile In My Shoes (by Jonathan Rundy)*, I talked about a brief encounter that I had with an unidentified flying object. I was about twelve years old. And it took place in the mid 1960s.

My job was to milk the cows in the morning before going to school. I frequently had to milk them again in the evening; usually when my parents failed to come home.

On this one night, I was walking behind the cows as I chased them along a creek to go to a shallow place to cross. Suddenly, I had the strangest sensation that I was being followed. So I stopped and looked behind me. That was when I saw a strange saucer shaped object hovering just a few feet above my head. I was, quite naturally, very scared. And took off running.

I regained my senses some three hours later. I was standing under a tree up near the barn. It was opposite to the direction that I had taken off in at a full gallop. To this day, I have no recollection of what may have transpired in those missing hours.

I have not retold this story for any reason other than I am unable to say whether that event had any influence on my mind. Did an alien alter my brain or give me knowledge? Who knows. It is one mystery we may never solve.

It is interesting to me and, hopefully, to you, too. That being said...

Most, if not all, great scientific achievements have been made as the result of a tremendous amount of trial and error. This trial and error process has been universally accepted as a necessary evil which cannot be circumvented in any appreciable, or erstwhile, manner.

Because of the very nature of the scientific process, a general rule of thumb has come to bear on the whole procedure. That is, we elect not to accept new thoughts or theorems without exhausting every avenue. Generally speaking.

If we have a question, we all know that there are numerous, albeit hypothetical, answers which crop up to explain the phenomenon. We attempt to eliminate the bad ones, eagerly accept the good ones, and anticipate alternatives.

The point is that we would be highly remiss if we chose a solution and disregarded any other possibility. So long as one possible, and contradictory, solution exists, we are obliged to pursue it doggedly to a final outcome. To do otherwise is a slap in the face to scientific advancement.

Why am I wasting the time and energy to explain things which you already know? Well, it would seem that you don't actually know these things. Huh?

One thing that has chagrined me in my quest to understand the universe around us is censorship by the elite. Their philosophy is that they have all the answers…well, at least, to their way of thinking, all the important answers.

I attempted to submit papers which would show that Albert Einstein was wrong (partially). I was ostracized and my papers discarded without even a cursory consideration. They, in fact, were never read.

How much more so would this censorship occur if they had known that I was refuting the great Mr. Newton? Discarded is discarded. But the accompanying laughter carries a certain magnitude which increases exponentially with the degree of lunacy attached to the event.

But am I the lunatic? Am I the one who has tuned out the truth, or possible truth, so that I can surround myself with my own vanity? Am I the one who will perform some scientific research, but outright reject any which might present a contradiction to my own beliefs?

Am I the one who envisioned microbes engaged in some fictitious form of tug-of-war with microscopic threads so as to hold celestial bodies in their respective orbits? Am I the one who said that there was a finite quantity of energy within an infinite system?

Did you hear me advance the notion that a blackhole is a place where gravity is so intense that nothing can escape it? Incidentally, there are numerous patrons, scattered hither and yon, who would have us believe that blackholes are the portals to parallel worlds, entrances to vast wormholes which instantly propel us to a position at the far end of the universe, and other fanciful musings!

Maybe I have the answers. Perhaps, not. But I have one thing that those others don't. I have the ability to engage in abject study without limiting myself to using "facts" handed down to me by a man who lived three hundred years ago. And I am willing to exhaust every avenue of research before making a scientific declaration of probability.

In my mind, the universe is not such a complicated place. There are logical answers to every mystery. I believe that I provide those within the pages of this book.

I know what gravity is. I know what blackholes really are. I know how and why the planets move. I know that the speed of light is near infinite in interstellar space. And I am on the verge of redefining the atom and electromagnetic radiations, in general.

So who am I? Do I work in the field in some official capacity? No. Do I have any college degree or official status of any kind? Absolutely not. Very few credentials of any kind exist in my past history.

So why should you waste your time reading any of the ramblings which I choose to ascribe to a blank page? Why not?

Science has answered a lot of questions. Still, many more have gone unanswered. That is why you picked up this book to begin with. You want to know; just as I wanted to know.

So, do I have the answers? I believe so. Only time will tell. But I can promise you two things. First, you won't be too terribly bored. Second, you will learn to challenge the accepted science (sic).

This book is as much about learning to think as it is about revolutionary new ideas. For instance, we have all been taught that gravity is a force of attraction. But what if no such force exists? What if gravity is about "push" and not about pull?

How does starlight travel so many light years and still reach us---intact???!!!

How come the speed of light is, allegedly, maxed out at 186,000 miles per second? Newton said that a body in motion will continue to do whatever it is doing, so long as it isn't being acted upon by an outside force. So, either light doesn't have a body (mass) or it is being acted upon by an outside force. Or is there another answer?

Einstein told us that there is an observable relationship between energy and matter which is contingent upon the position of the observer. The relationship between energy and matter is pretty much constant. Any differences are, therefore, contingent on the observer's observations which, in turn, are dictated by numerous other factors.

Incidentally, just so you'll know, I make several comments regarding the illustrious Albert Einstein. Throughout this book, I will continue this because I have studied the man's work, his life, and, particularly, the effect of that work on modern physics. I do this because, as far as the layman is concerned, he is the most well known of the men who shaped physics in the twentieth century.

I must apologize if it seems that I am ignoring, or somehow downplaying, the roles of other great men. Bohr's model of the atom, Heisenberg's Uncertainty principal, Mendeleev, Hertz, on and on, etc., etc. All of these things are important contributions.

Nonetheless, there is one very good reason why I opted to tout the work of Albert Einstein. You see, the man spent a sizeable portion of his life searching for a Unified Field theory. I believe I know why it eluded him.

Albert devoted so much time to defining the relationship of energy to matter, or matter to energy, that he all but ignored space. He got very close to defining gravity with the notion of an elastic space. And he was close when it came to the knowledge that a body in space causes a distortion in space which bends light, etc. So close.

Einstein made a blunder which so many others have perpetuated. He underestimated space. He separated space from both energy and matter.

The key to defining Unified Theory is not in defining energy/matter. The whole key is not in defining gravity. Nope. The whole key is in defining space as an energy which acts with a duality (much as light).

Duality? Does that mean that it behaves as both a particle and as a wave? In a manner of speaking, yes, you could say that. But there is more to it than that.

Duality? Does that mean that one part of space interacts with matter, while another part interacts with energy? You're getting warmer. In fact. You are boiling.

Let's leave it at that for the time being. Consider it a tantalizing tidbit which whets your appetite. But don't worry, you don't have to read the whole novel to solve the mystery. That would, after all, be a bit cruel. You can always skip ahead to chapter fourteen and work your way back.

Keep in mind that this book is compiled in no set format. In many cases, I am, literally, working on the theories at the same time I write. What I mean is, do not accept any singular answer as the right one. My intention was to share my thought processes as much as to share the results.

Science is trial and error. You get the benefit of my errors as well as my successes. Please keep this in mind as you read. Otherwise, you might dismiss me as a total quack and miss some very important revelations!

That being said, I wish to make a special dedication to all of the many thousands of men, and women, who have struggled throughout history to try to ascertain the precise nature of the world around us. We have not done so with any express desire to achieve fame or fortune (not that any of us would ever turn that down).

There is a certain amount of satisfaction that comes with comprehension. On the other side of the coin is the frustration of skirting the wrong paths in an attempt to see over the hedges. Let's all hope that we are finally on the right road to clearer vision.

NO FORCE OF ATTRACTION

By 2004, I had almost completely abandoned any notion that gravity could be a force of attraction. The reality had set in that no such force as attraction exists. Nor can such a force be observed anywhere in the cosmos. And, yet, the whole world continues to embrace it. Why?

The answer is simple enough. People prefer the easy way out. Instead of looking for complex fields, fields which behaved as waves pushing on any object within its path, it was easier to envision a force of attraction. And so, Newton's Law survives. So far.

The one thing that amazes me is the staunch refusal of Physicists to accept that which their own eyes reveal to them. Do we see planets banging into each other? Certainly not. In fact, we see just the opposite.

Man, for whatever reason, seems to like to make things more complicated than they really are. I do not know if this logic is a progression of those who prefer to think in abstract terms, or if it is just the belief that complex systems cannot have a simple explanation. Whatever the rationale, we continue to make simple events more complex.

The fact is, we live in a solar system which is part of a much larger galaxy. And that galaxy was the product of an immense explosion. No, not the much touted Big Bang. Something considerably smaller.

An explosion slows down, and even stops, particles within it. Our galaxy and, hence, our solar system, is acting in slow motion. Like it, or not, we live within a closed system in which time is not 'normal.'

What is 'normal' time? Well, if we were to exit our galaxy and enter into "true" space, we would encounter normal time. In statistical analysis, we can say that 'normal' is the state, or condition, inherent in the majority of our observations. All of the galaxies in the universe would only account for a small fraction of the total universe. Therefore, normal time is that time which exists outside of galaxies.

If we were at a point outside of our galaxy, somewhere in the vast expanse of interstellar "true" space, we might observe the lifespan of our galaxy as a mere singularity which might not be as much as a flicker of light that appears and disappears so rapidly that we would miss it if we occasioned to blink.

But we do not reside outside of our galaxy. We are in the midst of an event that created our solar system. This event is an explosion which created a bubble in which pressure is radiated inward from the outer walls of the galaxy. It is this pressure which provides us with the illusion that gravity is a force of attraction.

A lot of people do not fully understand the nature of an explosion. Essentially, an explosion is a complex field of thermal exchanges. By thinking in terms of thermodynamics, we can paint a very precise picture of the world around us.

But what is thermodynamics? What, exactly, is the difference between hot and cold? We know we get burned by very hot material. Many of us know that we, also, get burned by very cold material. In a sense, there does not appear to be any appreciable difference between extremely hot and extremely cold. Why?

Hot and cold, as we understand them, is merely a difference in the rate of speed of a given unit. That is to say, every molecule in the universe has a speed limit. It also acts in a relative manner. It makes little difference whether the molecule is going a thousand times faster, or a thousand times slower, than we are.

We see this phenomena everyday. It usually manifests as electricity. As long as we have two similar charges, at similar magnitudes, there will be no difference in the potential between them. But, as soon as one becomes greater than, or weaker than, the other, the difference in potential will cause a riff between them. This riff, generally speaking, will appear as a spark that leaps from greater charges to lower charges.

This is the simplistic view; the simple explanation. The reality is that there will be a difference in the rate of speed that one unit is traveling. And this difference in speed is what we observe.

Scientists are going to get balmy over that assessment. And that is alright. They just don't get it. They prefer to think that there is a mystical entity known as positive and/or negative. This mystical entity is standing there playing catch with other mystical entities.

The first rule of the Universe is that all things move. The atom is always moving. We are always moving. Everything is moving.

The second rule is that a given unit, be it the atom or a planet, is always going to take the path of least resistance. This means that bigger things will push littler things out of the way. It also means that littler things can move faster in small spaces than their huge contemporaries.

I do not wish to go into a lengthy discourse on motion, at this time, so let me give you a simple analogy that should make this concept more vivid in your gray matter.

Let us suppose that you are in your automobile. Let us suppose that your automobile is traveling down the highway at fifty miles an hour. Now, let us introduce a corner that is ninety degrees. You tap your brakes to slow down. But, alas, your brakes are nonexistent. What do you suppose happens next?

Your car wants to travel fifty miles an hour. It also wants to continue in a straight line because that is the path of least resistance (or so it thinks). You, obviously, fail to negotiate the ninety degree turn and your car sails off into no-man's land.

A man standing next to a huge oak tree watches as your car crashes into the tree. He assumes that the tree has some magical property which attracted your vehicle towards it. Is he wrong for thinking that the tree was magic, or are you wrong for thinking you could negotiate the turn at high rates of speed?

Additionally, you would be wrong if you have formulated the notion that gravity is the product of any particular motion. I, sometimes, make those spiels just to show you the conventional mode of thought. Indeed, I actually thought that myself. But that was long ago and far, far, away. Literally.

Gravity is a product of space. By the time you finish reading this book, you will have a very clear and precise knowledge of gravity. And you will see the universe more clearly than you could ever have imagined. And the simplicity of it all will make you laugh.

Is the Earth flat? Does the sun revolve around the blue planet? Gravity is a force of attraction? These questions are all redundant. And we will cry out in unison that the answer to each is a resounding, "no!"

SPACE COUNTS, TOO

Thanks to Einstein, not to mention his contemporaries, the focal point of the Physic's world seems to be centered around the relationship of energy to matter. In fact, I fell into that same trap and even considered that one might be converted into the other. It is certainly arguable. However, I have since come to the conclusion that we have sorely overlooked the most important of the three observables.

Our universe appears to consist of three distinctly separate variables. There is energy, matter, and space. We think we have learned to master matter. We have recently learned how to control many sources of energy by manipulation. And the Wright brothers proved that man can fly. But what about space?

Space is truly the final frontier. It surrounds us on all sides. And, yet, we know very, very, little about it. This is, after all, only logical.

We can see matter. We can see energy. We cannot see space (at least not in the colonial sense). And so it has been relegated to a lowly position on the grand scale of importance. Man likes convenience. Space is not at all convenient. It shields itself in darkness.

Energy and matter behave as they do only because space has made allowances for these things to occur. Man has reached the point where it is not enough to control energy and matter. We must drive ourselves to learn the inner workings of space. It is that knowledge which will propel us through the 21st century in an equitable fashion.

In order to figure out the functioning of space, man is going to have to first dispel any preconceived (mis)perceptions he has about the universe around us. For instance, instead of attributing planetary motions to some fictitious umbilical cord which tethers one to the other, we need to reassign that feature to space.

When I see the moon, or any planetary body which does not appear to be rotating on its axis, I do not envision that occurrence as a restriction caused by invisible tethering to the mother planet/host. What I see is a result that was caused by a constriction of space. From that vantage point, I begin comparisons to bodies which rotate.

Upon observation, one begins to suspect that celestial bodies revolve because they have atmospheres which act as lubricants against an abrasive space. However, some bodies, such as

Mars, do not have atmospheres. So what can be the influence or governor?

Polar cap(s) could be the simple solution. But I suspect that it may be a combination of factors. One clue comes from a gyroscope. Gyroscopes seem to display functions which its individual components would be incapable of. Therefore, we can deduce that a planetary unit revolves due to some inner workings. Specifically, liquid cores would be more active than solid, stationary, cores.

As I gaze heavenward, I begin to suspect things about space which most people would never give a second thought to. Or any thought, for that matter. Please excuse the pun.

One thing that I suspect is a sliding space. We see planets orbiting around a distant sun. Nothing strange, or remarkable, about that. Or is there?

A planet changes direction as it goes around the sun. We say that gravity from the sun has grabbed the planet in such a way that it whips it around and tosses it off in the opposite direction. What gets little attention is what is occurring on the other end of the orbit. What, precisely, is out there which is grabbing the planet and tossing it back towards the sun?

In the universe within the limited framework of my mind, I see a situation wherein space is enjoined to a body in motion. It is not just the planet which is getting slung out into space…it is also the space which has attached itself to the body which is getting slung.

I shall let you dwell on that thought while I attempt to put my universe on paper in such a way that it makes some modicum of sense. If I do it right, perhaps I can explain the relationship of energy to matter to space with a clarity deserving of a closer inspection by the people who have the means to perform those secular tasks. Perhaps.

LIGHT vs SPACE ?

As my feeble mind explored the universe around me, I began to realize that space was not just an empty region where things were able to move from one point to another. Space was, itself, a force. Not just any ol' force, but a force to be reckoned with.

I think a lot of us have taken space for granted. It has existed as something we cannot define. Nor can we adequately explain it in any appreciable way. Nonetheless, it sits there taunting us with silent mockery.

There can be no doubt that space is an active force. It keeps planets separated. It pushes planets together. It allows some to rotate; others can't. But what is the nature of this force?

I wish I could answer that question. Unfortunately, I can only entertain wild conjectures as I stumble across the galaxy. For instance, I am of the opinion that light and space may even be the same entity.

When I first advanced the notion that light and space were the same force, I was greeted with substantially more than repressed consternation. In fact, there were some of my friends, not to mention family, who ascertained that I was a prime candidate for the loony bin. Be that as it may, I continued to ponder the possibility.

Light in lieu of space makes for some rather interesting perceptions. We know that if we squeeze a sponge, we squeeze the air (space) out of it. We know that when we squeeze (apply pressure) on various solids, liquids, and gases, that we cause energy to be expelled from those entities in the form of light, electricity, heat, etc.

It is not hard to make the argument for the equivalence of light and space. What is hard is to get our minds to rethink the way in which we perceive these things. Is a cloudy answer not better than no answer at all?

One thing that becomes readily apparent is the distinct possibility that there exists two kinds of space. The first would be a space which interacts with all things…a kind of blackboard for creation. In the jargon of modern day Physicists, this would be the weak force space.

The other type of space would be the nuclear, or strong, force. This space only interacts with the weak force space. In other words, it does not, directly, interact with macroscopic particles. Huh?

Let me express this is less complex terminology (as if it could get any simpler). One force, the weak force, interacts with microscopic entities on a molecular level. The other force can only interact with large quantities of the weak force. Hence, the strong force only acts macroscopically but retains a certain magnitude of influence on lesser forces.

Many people who accept this concept try to make the distinctions more precise. Some say that the weak force only works on energy (microscopic force) and the strong force only works on matter. This rationale presupposes that there is a barrier between energy and matter which will not allow one to convert to the other. Can this be?

Absolutely. Energy is a small (relatively speaking) force, whereas matter is much larger. The two are so obviously different. But the difference is more complicated than mere size or speed.

One crucial difference between energy and matter is that matter cannot exist without energy, but energy can, very easily, exist without matter. This is a bit hard for some to fathom as we are all taught to believe in some symbiotic relationship between the two.

Take, for instance, the case of matter and antimatter. In reality, matter is a three dimensional entity and antimatter would be a three dimensional void. Irregardless, we still treat each as an energy form. Why?

On some subconscious level, we understand that the mere presence of matter represents the presence of energy. Matter, whether in motion or at a standstill, reminds us that it takes a certain amount of force to move it. So the temptation is to assume that matter, itself, is a force. The reality is quite different.

Remembering that matter cannot exist independent of energy, we can deduce that it takes X amount of force to move matter because matter has X amount of force holding it in place. No other reason. Matter, therefore, has no force of its own.

Going back to the argument that the weak force works on energy while the strong force works on matter, can we accept that rationale. The outcome is that this appears so. Nonetheless, adhering to the preceding argument, we must conclude that the weak force interacts with matter while the strong force interacts only with significant concentrations of the weak force.

Does my assertion that space and light are the same entity bear any credence? Yes...but only within the confines of a closed system. In the vast expanse of interstellar space, this changes.

PLANETARY MOTION

When I wrote my book on Unified Field, I made a minor blunder which actually led to my current thinking. You see, I looked at the Planetary Pairing in our solar system and derived at a formula that I stated as $M_a + M_b = M_c + M_d$. My mistake was that I was really using the planets' diameters and the formula should have been $d_1 + d_2 = d_3 + d_4$.

For your information, the above formula is one I formulated to calculate the relationship of each planet in a pair. d_1 is the diameter of the first planet. d_2 is the diameter of the first planet's moon(s). d_3 is the diameter of the second planet in the pair. d_4 is the diameter of the moon(s) of the second planet.

Things always seem to work out for the better. And this is in spite of how badly we, ourselves, may goof. By substituting mass for diameter, I really paved the way for Unified Field. Huh?

When we are dealing with very small bodies, such as those found in atoms, we compare physical size with considerable regard for density. However, when we reach cosmic, or macrocosmic, size, we can ignore density and must then consider only the total diameter (spatially speaking). That is to say, we must consider the size of each planet and its various constituents.

What I was overlooking was the precise way in which forces interact. I had not carried the hypothesis to its logical conclusion. And, so, I needed to finish that which I had started. Let's start with a planet's orbit around the sun.

When I examined the statistics regarding the speed in which a planet revolves around the sun, I discovered an interesting equation. The closer the planet was to the sun, the faster it moved. If the planet were tethered to the sun, in any appreciable way, then one would expect the opposite. Obviously, the planets had no such tethering.

My next thought was that the nearer an object got to an energy source, in this case the sun, the greater the effect of the energy. Seems logical, doesn't it? Guess again.

If the resulting burst of speed were the result of an energy increase, one would, also, expect to see a similar increase in the speed in which a planet rotates upon its axis. But that just doesn't happen.

If we eliminate the energy of the sun as being the cause of the acceleration of the planet at perihelion, then we need to look at space as being the responsible party. Is this plausible?

According to Einstein's postulates, space is compacted around a body. What this means is that space becomes denser the closer we come to a body such as the sun. Interesting.

If we run with that simple logic, then it would stand to reason that any planet approaching denser space would be forced to slow down as it would require more energy in order to propel through denser space. However, this additional energy is made available by its proximity to the source.

By my estimation, the denser space couples with the increased energy in such a way that a planet's journey around the sun would be at a constant rate. This, however, is disputed by Kepler's laws of planetary motion...as well as by observation.

So what is the explanation for this seeming paradox? We haven't considered all of the forces at work. Remember, any motion is the net result of all forces. In this instance, we have neglected to consider the sun's motion.

The sun rotates on its axis at the rate of approximately one revolution to every twenty-seven earth days. The dense space near the sun moves at a faster linear rate (parallel to the surface of the sun) than the less dense space fartherest from the sun.

A proper analogy would be to think of the sun as a flywheel and each planet as a pressure plate. Space, itself, would be the clutch plate. As the planet moves closer to the sun, the greater the pressure on the clutch plate (space). This causes the pressure plate (planet) to try to match the flywheel speed (27 days). And the planet speeds up.

For all intents and purposes, the sideways motion of planets is the result of space. If the sun were to stop rotating, the planets would stop revolving around it.

Our next problem comes from a planet's rotation upon its axis. Any motion, as we are all well aware, is the result of (net) forces. This would imply that the more a planet rotates, the greater the forces acting upon it. In this sense, we would expect the larger planets to rotate upon their axis at a greater rate of speed than their smaller counterparts. We would, also, expect that this rate would be influenced by a closer proximity to the source (sun).

Now, many of you are going to oppose this theory based on the motions of Mercury and Venus. And well you should. You see, these two planets rotate on their axis at a far slower rate than any other planet and, yet, they are the two closest to the sun!

For the solution to this little dilemma, we first turn to the obvious. Are planets' rates of rotation determined by atmosphere? Since Mercury has no atmosphere, we would expect it not to rotate very fast. However, Venus has an atmosphere. And Venus rotates very, very, slowly.

The atmosphere thing seemed like such a very logical conclusion. This is particularly true when one thinks in terms of ball bearings (friction). Obviously, a ball bearing which has little, or no, lubricant, will rotate less freely than one that has lubrication. Venus defies this.

And so does Mars. While Mars rotates at approximately the same rate of speed as the Earth, it has, virtually, no atmosphere. So what is the answer?

The problem of planetary rotation seems to lie in two factors. First, the presence, or lack of presence, of polar caps. Secondly, the existence of thermal variances.

In order for anything to rotate, uniformly, it needs a focal point, or fulcrum. This is solved by magnetic poles which align themselves with the magnetic field induced by energy passing through space.

Unlike electricity, magnetism is a "cold" force. Thanks to this thermal property, ice forms at the poles. An interesting property of ice is that as it transmutes from a liquid into a solid, it expands. This expansion makes the space, it occupies, less dense. This, in turn, means that it takes less energy to make the planet rotate.

The stage is now set. All we need is an energy source which will cause the planet to spin. This is where the temperature variance comes into play.

Most people would assume, if they give it any thought at all, that planets rotate because they are asymmetrical. The logic is that either solar wind, or sunlight, pushes on the planet's surface. Since one side of the planet (left or right) is lighter than the other side, this pressure forces the planet to spin. This is erroneous.

Let me give you an analogy. Let's say that I have two one gallon cans and I fill each with gasoline. In the first can, I pour gas that has been preheated to one hundred degrees. In the other can, I pour gas that has been cooled down to thirty degrees. Which one will propel my car the furtherest?

When I heated the gasoline up, it expanded. This means that, on the molecular level, the heated gasoline contains far less molecules of gas than the cooler quantity. Hence, I shall get more energy out of a gallon of cold gas than I will out of a gallon of hot gas.

Now let us return to our planet. As the sunlight heats the surface of the planet, it expands the molecules and, like the polar caps, makes space less dense. This enables the planet to rotate. How?

As the "atmosphere" heats up, the heat expands to the point that it seeps around the planet and encounters the cold, dark, side. This heat causes the cold to release energy as it expands. This release of energy is in the direction away from the cold. This, literally, pushes the planet and it rotates in the direction of the lesser force.

One thing that immediately springs to mind is that the greater the disparity between the hot and cold, the greater the rate of speed. And this is not contingent of planet size. What?

Yes, I know. You think that the bigger planets are rotating faster because there is more atmosphere (surface) and that represents an increase in energy and potential energy. But this is illusionary.

As the surface increases, there is a corresponding increase in energy. But there is, also, a corresponding increase in mass (this will be explained in my section on gravity). The increase in mass cancels out the gains in surface energy.

Have I lost my mind? How can I say that the increase in mass has nothing to do with the rate of speed when we can clearly see that the largest planets are rotating at a far greater rate of speed than their smaller counterparts?

Elementary. The increase in mass does, in fact, play a part. The increase in mass means an increase in the planet's magnetic field. This, in turn, means a much larger polar cap. To put it succinctly, the larger polar cap means more space "lubricant" and our "ball bearing" can now spin faster.

Pretty simple, huh?

And now I am going to contradict myself. Size does matter. It just doesn't matter for the reason(s) that you supposed it matters.

When I first met David McDaniel, he asked me, what he thought was, a trick question. The question he sought an answer to was: Does the outside of a given shaft rotate at the same rate as the center point? He was, of course, referring to revolutions per minute.

When I stated that the outside of the shaft rotated at a different rate, David and associates felt that they had tricked me into providing the wrong answer. They were sure that I was contemplating speed rather than revolutions. I wasn't.

The problem is a pretty basic one for engineers and involves torque created by disparities in acceleration. The center of the shaft requires less energy because it travels less distance. As we increase distance, we obviously need more energy to make the trip. The application of this energy is done in discreet packets and requires a certain amount of time to apply each packet of energy. In effect, the difference in acceleration creates a rotational torque which causes the outside of the shaft to travel at a different rate (revolution) than the center.

Rotational torque is usually lagging behind the application point. Sometimes, under the right conditions, it can become a leading force. In either event, it fluctuates because of tensile properties. And this creates a situation where the outside of the shaft is rarely rotating at the same rate as the center.

Incidentally, this difference in torque (not just friction) is why shafts, gears, etc., tend to get a little warm.

So what does any of this have to do with planetary motion? Think of each planet as being a shaft. The center rotates at a different rate than the surface. David's question made me aware that the planets all rotate at the same rate of rotation. Because the outside of the shaft has to travel further, it has a higher velocity than the center. And this gives us the illusion that the larger planets rotate faster than the smaller.

In all fairness to David, he is a very intellectual fella who enjoys contemplation of complex ideas. His lucid and ambitious thoughts have made conversations enjoyable, stimulating, and substantially impactive. To be succinct, Dave is as fine an intellect as you'll find anywhere. And I owe him a debt of gratitude.

Moving on, our next question seems to be: Is the point of application interior, or exterior, to the surface of the Earth and other planets? It is an excellent question and we shall endeavor to develop the answer in our next chapter. And we will return to the subject of planetary motion. And, still later in this book, I will reveal the truth about planetary motion, magnetism, etc. I did, after all, say that I was going to get you thinking.

THE COSMIC CONSTANT

To calculate the origin of the force which compels planetary motion, I needed to define that force. To be very specific, I needed a Cosmic Constant. That is to say, I needed to derive at a formula which would account for any motion presented by a large body in motion.

Two possibilities occurred which could not be ruled out. First, if sunlight is the primary force, it is very possible that the heating of the planet's atmosphere, or surface, could result in a motion laterally to the force represented by space.

The second possibility was evident in the fact that the sun's energy could be channeled into the core of each planet. That energy could be coupled with the energy present because of tremendous pressure exerted by a planet's mass on the core.

In order to eliminate one of the theories, perhaps even both, we needed to better understand a planet's movements. To do this, I engaged the notion that there is a cosmic constant which dictates planetary motion. What I mean is that there is a speed limit on all bodies which rotate in space (We shall deal with the problem of a planet's motion around the sun at a later time).

In metallurgy we encounter a very old process for making lead shot. Essentially, we take a crucible filled with molten metal and pour it into a vat of water. The molten metal forms little balls which solidify upon contact with the cooler water.

If a planet were a molten batch of matter at its inception, we would expect the planet to cool down into a ball. Not so hard to envision.

As our molten batch of matter is moving through space, it is revolving. During the cooling down process, the heavier material is condensed into the core. The lighter material floats to the surface.

One thing we know about physics is that as we pile, or increase, the amount of matter that we place on a given point, the greater the pressure on that point. And the more pressure there is on a point, the more heat that is generated at that point.

What I am getting at is that even if there were no molten ball of matter in the beginning of the creation of a planet, the pressure of the matter that makes up the planet would heat up the core as its mass increased (by accumulation of space debris, asteroids, etc.). The heaviest material, presumably metal, will always go to the center.

It is important to remember that a planet's core is made up of its heaviest material. This heavy core becomes the fulcrum for the planet to rotate along. It provides the planet with a certain amount of stability…and independence.

Under this postulate, any planet which has little, or no, metal in its core, cannot rotate freely upon its (nonexistent) axis. Thus, we would expect that Mercury, Venus, and our moon, must not have a concentration of metal at their cores.

Notice that I insisted that the metal be in the core. This is not the same thing as saying that these bodies have any less metal than their contemporaries. Balls with heavy cores move differently than balls with light or hollow centers.

So what, if anything, does any of this have to do with a cosmically constant rate of rotation for heavenly bodies? Well, as we attempt to formulate an equation which best defines this feature, we need to be mindful of the factors which hinder and/or help the process.

Another important variable is the thermodynamic opposites presented by each. For simplicity, let us assign each a digital signature. By this, I mean, let us say that a given region is either hot (on) or cold (off). This is fairly simple (until we get into the regions which border the transition from hot to cold).

Be that as it may, let us return to the notion of a simplistic Cosmic Constant. What variable can we use which will enable us to define the forces at work around us?

In a word: Gravity. But how can we use gravity as a Cosmic Constant when we cannot even define what it is? Elementary, my dear Watson. Elementary.

Many Physicists presume that gravity is a consequence of motion because it seems to imitate an accelerating force. I, myself, fell into that trap as I considered the speeds at which Earth moves. I.E., it rotates on its axis and has an equatorial speed of about one thousand miles per hour and zips around the sun at an astonishing 66,000 miles per hour.

It is pretty easy to see why somebody would suspect that motion played an integral part in the formulation of gravity. Therefore it becomes rather ironic that speed has nothing, whatsoever, to do with gravity.

Gravity is a function of space which is, technically, independent of either matter or energy. Space is the medium which houses both energy and matter. Indeed, to reiterate my earlier ramblings, a unit of space is assigned to every unit of matter.

I have heard it said that Newton's theories regarding gravitational attraction (sic) reveal that it makes no difference what a body is made of. Its size is all that matters. In fact, Newton would have believed that the weight of an object would mandate that the object possessed more gravity the heavier it got. And then he contemplated placing heavy objects and light objects within a vacuum. Oh my.

And there the puzzle remained. If weight made no difference, and size did not seem to make any difference, in a vacuum, what in the world could be causing gravitational attraction? It was a point that Einstein would later muse over.

By the time Einstein developed his theories of Relativity, he had given up on the idea of space as an interacting entity. That is, Albert decided that space, specifically the Ether, did not have anything to do with Relativity.

Einstein would spend the greater portion of his life searching for an elusive "Unified Field Theory." He postulated that there should be some formula, or two, which would define the universe. How sad that he gave up on the Ether's importance.

My formula for ascertaining gravity is radius times density. That is very simple and straightforward. And perfectly defines gravity. And it is the basis for the Unified Field theory that I will give you soon.

Gravity is a function of space, and space alone. The size of the body and its density is what creates gravity. Let me give you a very simple analogy which will draw this in your mind.

Let us take a simple sponge. We throw it in a bowl of water and it immediately soaks it up. When the sponge is saturated, we wring it out and it is, once again, ready to absorb water.

Now think of that ball of dirt in space. At the molecular level, that ball of dirt has many units of space within it. If we wring that ball of dirt out by compressing it, all of those units of space rise to the surface of the ball of dirt. The more we compress the ball of dirt, the more space molecules rise up and congregate around the surface.

Remember that all of that displaced space is energy. It is energy that is both pressing on the surface of the planet and against galactic space. More than that, it is a congregation of weak energy which can now be affected by the movement of the stronger galactic force.

Einstein provided us with the relationship of energy to matter which he defined as $E=MC^2$. The famous equation provided seemingly simple solutions to many questions which existed at the time. Many professionals interpreted the formula to mean that energy could be converted to matter and vice versa. Not so.

Certain restrictions exist. First and foremost, you can't change a dog into a cat. Secondly, you cannot accelerate matter to the speed of light. This latter fact has been explained away by stating that it would take infinite energy to achieve the feat. This is an over-simplification which is grossly erroneous.

The cosmic constant says that there exists a given quantity of energy, and space, for every unit of matter. I leave it to the pros to state this in numerically exact terms, but proffer that you can neither add to, nor take away from, that constant. In essence, this defines the Laws of Conservation. The question, at least in my mind, can the laws be violated?

Envision, if you will, a huge ball of material. Let us call this Baby K. And let us suppose that it contains every bit of energy and matter within our galaxy.

Now let us ignite Baby K (how we do this will be fuel for further discussion in another time and place). What happens?

As Baby K disintegrates into smaller bodies, the weak force (gravity) particles that once surrounded the huge ball of material are now being used as "filler" or space. Gravity molecules are scattered everywhere and, yet, must remain within the system.

Large chunks of cosmic debris have large amounts of gravity molecules. A uniformity exists as weak space (gravity) combines with matter and matter's companion, energy.

What can we conclude from this exercise? First, we exist within the closed system of the Milky Way. In particular, we exist within the closed system of our solar system. Second, have a very exact amount of energy, a very exact amount of matter, and a very exact amount of weak force space. Hence, energy and matter are conserved.

But what about the weak force space? Again, it, too, is conserved. It is conserved by virtue of the fact that it coexists with the energy/matter material. It becomes a precise coefficient of same.

One word of warning. Do not get weak force space (gravity) mixed up with strong force space. While weak force is limited, strong force space is infinite. And behaves quite differently than weak force space.

Something to think about.

PLANETARY PAIRING

People are hard-pressed to dispute my contention that the planets are paired up. After all, you can see those things with your own eyes. And facts are very hard to circumvent…no matter how clever one may approach the attempt.

Uranus is obviously paired up with Neptune. Both are approximately 50,000 km in diameter. And both rotate on their axis at a rate of approximately 17 Earth hours (making the Uranus and Neptune days 17 hours each).

The largest discrepancy in the solar system pairing is that between Saturn and Jupiter. There is a very logical explanation for this, and that will be addressed shortly.

Saturn and Jupiter are mates. They are the two largest planets in our system and that, alone, dictates that they are paired. Indeed, they are so closely matched that it is inescapable.

Saturn is 119,871km (roughly 74,500 miles) in diameter. Saturn rotates upon its axis at the rate of approximately 10.67 earth hours. Compare those stats with Jupiter...weighing in at 142,800km (88,736 miles) in diameter and a day comprised of some 9.8 earth hours. Very, very close. Too close to be coincidence.

Our next pair would be Earth and Venus. Earth is 12,753km (7,926 miles) in diameter while Venus is 12,104km (7,522 miles). Obvious mates.

Mars weighs in at 6,785km (4,217 miles) and Mercury supports 4,878km (3,030 miles) in diameter. Just for comparison, the Earth's moon is 3,474km in diameter (approximately one-quarter the size of the Earth itself).

The first thing we notice about the pairing up of the inner four (or smallest) planets is that they are not, necessarily, neighbors. Venus and Earth are neighbors, but Mercury and Mars are not.

The next discrepancy comes with the fact that no pair of the smallest planets rotate at the same speed (as the larger ones do). That is to say, The Earth/Venus pair are dissimilar. Likewise for the Mercury/Mars pair.

The question arises, why the discrepancies? The answer is fairly easy when the whole solar system is taken into account. And, surprisingly, different from that explanation offered by professionals who insist that Venus and Mercury rotate more slowly because of the sun's immense gravitational pull.

We see an Asteroid Belt which is in between the four smallest planets and the four largest. We are all cognizant of the so-called Roche limit. I think it a safe assumption that there is an invisible barrier which crushes large planetoids into rubble.

We could, rather easily make an argument for the obvious. Namely, we could argue that the four inner planets were, at some period in their lives, moons of the four larger planets.

But let's not go that far back. Instead, let us continue our examination of the discrepancies.

The next interesting fact to rear its head in mockery is the rate of rotation for Mars. You see, Mars rotates on its axis at the rate of about 24.6 earth hours. That means that the Martian day is not so very different from an average Earth day (in terms of length).

Now wait a minute. It is very obvious to even the most illiterate amongst us that the larger a body is, the faster it rotates. Conversely, the smaller a planetary body is, the slower it rotates. So why is it that a planet that is so much smaller than the Earth is rotating at a speed comparable to the Earth?

Several variations can be concocted about the scenario I am about to describe. However, the gist of what I am saying will shed considerable light on the subject.

At some point, either the Earth, or Venus, was orbiting where Mars currently resides. My best guess is that it was Venus. Irregardless, under the premise of planetary pairing, Venus and Earth each had one moon and both rotated at about twenty-four earth hours.

Along comes a large celestial body (possibly even Mars). This celestial body slams into Venus and pushes Venus out of its orbit. As Venus heads towards the sun it collides with its moon, Mercury, and Mercury heads towards the sun, too.

At some point in their journey, either Mercury or Venus, probably mercury, strikes the earth moon. Mercury veers off at a different trajectory than Venus and each of them assumes their new place within the solar system.

According to Newton, whatever a body is doing, it will continue to do until acted upon by an outside force. This is important to the whole picture.

When we take two spinning objects and collide them, there is a tendency for two events to occur. The first event is that the objects will change their initial direction of travel. The second is that they will slow their spinning, wobble, or even stop.

All things in the universe spin. However, in the event of a collision, this is subject to change. Nothing so mysterious there.

Moving on, the problem that people seem to have is why are the planets paired up? And, I have to admit, I was at a bit of a loss to explain that, myself. And so it was that I opted to explore some avenues of possibility.

Ultimately, I came to the conclusion that Planetary Pairing occurs as a defense, or survival, mechanism. When one planet meets disaster, or gets pushed out of its normal flight path, the system allows a new planet to occupy the space of the former occupant. The only rule that seems to apply to this is that the new planet has to be the same size, or smaller. The new planet cannot be bigger than its predecessor.

An interesting side excursion to this postulate is the concept of a shrinking universe. Or, rather, a shrinking solar system. If the system can only accept the same size, or smaller, then anything larger gets the boot. This leaves us with only one direction of growth. Or does it?

Now here comes a trick question…so be careful. If planets only spin at the same approximate rate when they are of the same, or similar, size, why is it that Mars and Earth are both rotating on their axis at about twenty-four hours? Something is amiss.

When Venus was knocked out of its orbit, Mars took its place. Did you forget Newton? The space, where Venus used to be, continued to spin at the same rate as if Venus was still there. Hence, Mars is rotating at about twenty-four hours per day because of spatial qualities rather than its own mass.

Interestingly, if this is true, then Mars should be slowing down. And you are wondering why it hasn't slowed down already? The calamity that caused the disruption of the Earth-Venus pairing did not occur that long ago. So, on the cosmic scale, the event has not yet played out.

Proof? Well, one only has to look at Venus. See how proactive the planet is? Envision the planet as having originally started out as Earth's true twin. Oceans, rivers, creeks, air that was compatible with life as we know it.

If the event had occurred significantly long ago, Venus would more closely resemble Mercury. For all intents and purposes, Venus is dying. But, then, you already knew that!

DUAL MASS THEORY

I have always said that the way to discover how atoms work is to take a good look at the solar system. So I listened to myself. I took a long, hard, look at it. And I was shocked at what I discovered.

Planets were not content just to have moons (or not). They had to have companion planets. That's right. Excluding Mercury and Venus, every planet in the solar system has moons. And these planets also have a twin (in a manner of speaking).

I know that I covered this in the previous chapter. However, there are some things worth repeating. And I think this is one of those things. It is just too obvious to be a coincidence.

Ever take a good look at the statistics? Earth rotates on its axis once every 24 hours and 4 minutes. Mars rotates on its axis every 24 hours and 37 minutes. Pretty close!

Jupiter rotates once every 9 hours and 55 minutes. Saturn does it in 10 hours and 39 minutes.

Uranus is 17 hours and 14 minutes. While Neptune is 16 hours and 7 minutes.

Most scientists and researchers, myself included, tend not to pay a whole lot of attention to Pluto. It, after all, looks more like a pair of moons. But who am I to argue. If it is a planet, fine. That means that it is the best example of this planetary pairing in that it flip flops with Charon, its so-called moon. So that makes it the most clear-cut example of planetary pairing.

Before launching into a tirade about why the others are all paired, let me offer an explanation as to why Venus and Mercury seem to be odd balls.

Venus is only fractionally smaller in diameter than the Earth. This is noteworthy because all of the other pairs are very close to the same diameter. More on that in a minute.

Mercury is only slightly bigger than Earth's moon. This leads me to believe that Mercury is Venus' moon. I suspect that if Mercury were orbiting Venus, then Venus, and not Mars, would be Earth's twin. And Venus would be rotating on its axis at pretty close to 24 hours instead of 243 days.

Is there any data which would support this theory? Oodles. Take the following calculations, for instance. The diameter of the Earth is about 12,756 km. The diameter of the Earth's moon is approximately 3,476 km. That adds up to 16,232 km. Right now, Mars and its two moons are about half the size of the Earth (alone). So we know that

Mars is not the twin for Earth. So let us look at Venus.

Venus has a diameter of about 12,103 km. Mercury has a diameter of approximately 4,878 km. The total for those two objects is 16,981 km. Take away the Earth/moon total and we have a difference of only 749 km. Pretty darn close!

For the record, the margin of error can be resolved by assuming that the precise measurements have been miscalculated and/or their respective densities are inaccurate. Alternately, the orbits of their moons might make up the difference in total mass/energy (though this is not likely as gravity seems to be independent of motion).

A third possibility may be that, in pairing, one of the planetary pairs has to be slightly smaller so as to enable one to lead and the other to follow. More precise calculations of the entire solar system (pairs) would probably reveal this...if the discrepancy is by the same amount in every pairing.

How would this Earth-moon/Venus-Mercury pairing compare to the other planetary pairs? Let us look...keeping in mind that Mars is an imposter. Mars, like Mercury, is a moon. And it belongs to the asteroid belt.

Another note, before examining the other groups, we can calculate the total mass contained in the asteroid belt and divide by two. This would give us a close approximation for the two planets (more than likely moons of Jupiter or Saturn) which collided...or were, somehow, smashed by an incoming immense mass). I propose that the larger planet be named Beverly and the twin named Baby K.

Jupiter has a diameter of about 142,980 km. It has approximately 16 (substantial) moons which I shall not even attempt to calculate for mass because there is a much easier way to calculate which planets should be paired up...as a rough estimate of little consequence.

Because of the angular velocity, or nature of light, it causes a thermal response on each planet in space. This response seems to be absolute in that each planet rotates on its axis in about 24 hrs or less. This allows us to calculate which planets are paired up (or should be).

Jupiter has 16 moons and rotates at just under ten hours. Saturn, diameter 120,540 km, has 18 moons and rotates at just over ten hours. Obvious mates. Ditto for Uranus, 51,120 km, and its 15 moons versus Neptune, 49,530 km, with its 8 moons. Pluto and Charon are similarly matched to each other.

Getting back to Mercury, let us assume, for a brief interlude, that Mercury gets repulsed from the sun. It may be trying to do this now. And that would certainly account for the little abnormality at aphelion.

Anyway, Mercury flies off and is captured by Venus (flies off because of the path torn in space created by the Sun when it entered our system). Total size, for moon and planet Venus, is now very nearly identical to Earth and its moon. And space is so structured that each level away from the sun needs two identical diameters per "ring" or layer. Hence, little ol' Mars gets shoved back into the asteroid belt.

Meanwhile, back at planet Earth, the Earth pole's reverse and Earth begins to rotate backwards. More than likely, this reversal will occur as a result of the Earth flipping over (so that it is continually spinning as it reverses).

Venus, likewise, reverses its poles and it, too begins spinning opposite to what it is now doing. The solar system is now balanced because Mercury/Venus is the right size, whereas Mars/Deimos-Phobos is slightly askew.

By the way, the date that Mercury rejoins Venus, as well as the date that the Earth's poles reverse, should be precisely calculated by observing the stutter step of Mercury at Aphelion. Compare length and magnitude. Since Mercury revolves around the sun every 88 days or so, it should be easy enough to do.

Continuing: The base formula for planetary calculations is going to be different than for the atom. The equation for the atom is $rd_1 + rd_2 = rd_3 + rd_4$. The r is for radius and the d is for density. Therefore, rd_1 is the measurement of the first proton in a pair, rd_2 is the measurement of the proton's electron(s), rd_3 is the second proton, and rd_4 is the second proton's electron(s).

In a very literal, and real, sense, this is the base formula for the Unified Field. While the formula appears to concern itself with mass, as a spatial quality, I am not entirely convinced and feel that it is equally important in other ways. For instance, by now, you know that radius times density is the formula for gravity. It is obvious, from that standpoint, that rd defines the size (spatial quality) as well as the weak force gravity.

Einstein's famous equation, $E=MC^2$, should come in handy for determining the energy to mass ratio. In that case, my Dual Mass equation becomes an energy calculator. My personal belief is that there is a direct correlation between energy and matter which states that it takes X amount of energy to push X amount of matter together. And this, in turn, will prove that energy and matter are *not* the same.

Bohr's model of the atom is obsolete. Niel's was a remarkably insightful, and intuitive man. All the more so for having made his postulates in an industrial stone-age! And I hope that people never forget the importance of his contribution.

In time, we will come to realize that the neutron, the proton, and the electron, are all units representative of mass, energy, and space. We can now calculate those forces by calculating the size of each component.

I reiterate that, by looking at what is going on in the solar system, we can determine what is going on at the subatomic level. It is the same force.

Several years ago, I developed a theory about electron pairing. Even if the formula was not so appealing to many, the name is marvelous. It has a nice ring to it. And it is possible.

Going back to the concept of Planetary Pairing, we might well want to reexamine the Dual Aspect theory of light. If we are reluctant to accept the dual atom as being the cause of the disparity, it is easy for us to utilize a dual photon postulate. They are small enough, and fast enough, that two of them could sail through a tiny slit. And when we opened the second slit, the faster second one would fly off, tangentially, and enter the second slot. Piece of cake.

I could go on and on and on. And I may, someday. Right now, I just want to finish this book, get it published and watch the fur fly. It will be fun to watch as hundreds of variants, calculations, and the like, spring forth from my groundwork.

I, for one, am excited to see if anything I may have missed comes to fruition. I predict something as important as the atom bomb was compared to relativity. Einstein could never have imagined such a thing coming to life in his own era.

I have always considered jet engines to be dinosaurs. As we progress, scientists will realize that space is a crosshatch of lines of force. By studying the cellular levels of the planetary pairs, they should be able to discover how these lines of force operate. Couple that with a more accurate depiction of atomic structure and, well, the sky is no longer the limit.

Cars will make jet propulsion obsolete by tapping into these lines of force. Hover cars? Probably sooner than later.

Isn't it nice to have something to look forward to? Maybe we all will get into the habit of climbing out of bed and reading the paper to see what new discoveries science has made overnight. I know I am.

ENERGY TO MASS RATIOS

Any kid in possession of a radiometer can tell you the effect that light has on matter. People with solar panels will hasten to support that erstwhile conclusion. The question then arises: why haven't we figured out a formula for calculating the precise amount of sunlight that it takes to move a given unit of mass?

Under my theory that planets need "caps" in order to rotate on their axis, I decided to use five planets as a "means" of calculating such a formula. I opted to use the Earth, Jupiter, Saturn, Uranus, and Neptune, as my focal group. I chose these planets because they all have the most commonalities. I.E.- atmospheres, polar caps, spin rates, etc.

I had two ways of approaching the problem. First, I could consider the planet's surface area. This seemed the most logical in that a collector could only gather up that amount of energy. In other words, that set the parameters for "flow."

As in the case of the solar panel system, we next needed a storage device for all the energy we harnessed. Obviously, the second consideration would be a planet's mass...viewing it as a potential storage unit.

Faced with an overabundance of information with which to calculate, I decided to simplify the procedure by eliminating variables which I concluded were, possibly, irrelevant. Further research may prove this wrong, but that leaves the door open for others to build on.

I elected to deal only with the net force. I should say, the apparent net force. In this way, I could ignore the material of both the collector (or accumulator, if you wish) and that of the storage mechanism.

Now many of you will not understand how I could ignore these two factors. I, myself, question the rationality of the decision. However, it is apparent to me that I need to concentrate on the end result. And what might that be?

Planetary motion does not occur because planets got bored and opted to dance their way across the universe. Planetary motion occurs because a planet has acquired an excess of energy and needs to dispel it. And so it moves.

The principle of Conservation of Energy does not allow us to simply state that the Sun's energy has merely vanished. It has to go somewhere. And

I do not subscribe to the Einsteinian notion that the energy is converted into matter.

Under my premise, a planet is saturated and that triggers motion. That seems like a logical conclusion. However, we must also consider the Law of Survival.

Under the law of Survival, an organism will do whatever is necessary to ensure that organism's survival. If a planet needs more energy than it is receiving, then it would become logical for it to increase input by rotating and moving around its source.

You may well have thought that through and asked yourself, why doesn't the planet move closer to the Sun so as to garner maximum energy? A lot of answers come to mind, including the idea that such a move would result in too much energy and doom the planet.

The Sun has a defense mechanism(s) which would not allow an onslaught. Ditto for the planets, but more pronounced with very hot objects with huge masses such as the sun's.

We have a solar wind of about 250km per hour. Do you have any idea how intense that is? Have you ever stuck your head out of the car at about sixty miles per hour? All kinds of forces act upon it. And, if you weren't that brave, you still sampled that force when you stuck your hand out.

Point is, you encountered a lot of force that was pushing against you in the opposite direction to that in which the car was traveling. We conclude that there exists some kind of relationship between energy (force) and matter.

We now have two constituents of the atom. We have the electron, which is force, and we have the proton, which is matter. Do you see the comparison?

Let's carry that postulate a bit further and assign each element of the atom a charge. We shall call the electron, negative, and the proton, positive. We all know that opposites attract and so we conclude that the electron is attracted to the proton. What keeps these two from banging into each other and canceling out?

Let's add the emulsifier. Let's add the lowly neutron. Where do we place it? And what, if anything, is it's precise function?

In classical quantum physics, we would tend to think of the neutron as a unit of mass. However, conventional thought is not much help when ascertaining the precise role of the neutron.

The neutron (sic) acts like a cocoon which envelopes the proton. The neutron, in that sense, is really a unit of weak-force space. Remember my earlier declaration that weak-force space interacts with both energy and matter? Well, what I didn't tell you is that this force acts equally on each. That is to say, space acts in the same magnitude for both energy and matter.

Another thing I did not mention was my belief that the electron strikes the proton, periodically, and that action causes the proton to rotate. Think of it as a lightning bolt which hits the proton on one side and that pushes that side of the proton, causing it to spin.

Now, as the proton spins, centrifugal force is causing the neutron to expand in a direction away from the proton. This pushes the electron further out and that makes it difficult for the electron to strike the proton. Hence, the rotation slows, the neutron condenses, and the electron can, then, repeat it's striking action.

In a sense, this same procedure is at work on any particle of matter. And this explains the Brownian motion we observe within a body of liquid.

Moreover, this perfectly explains the excitation of the atom and the flight of electrons to so-called outer shells. Does this mean that I do not believe in the existence of these shell layers?

To clarify that, let me draw your attention to solar system. In our system, the planets are spaced at very precise intervals. This is because of the tidal action of the energy radiating from the sun.

The atom tries to mimic the tidal action of the sun. But two important factors emerge which says that this cannot exist. First, the atom, or most atoms, are not radiating any energy of their own.

The second reason, and the most important, is simply that every atom we encounter is compacted by the mere presence of other atoms. In other words, there just isn't enough room (space) for shells to exist.

Consider, also, the charge. I abhor that word. I do not know, concisely why I hate the word; I just do. Reckon it has to do with my perceptions of charge.

What is a charge? In my mind, it is a direction of travel. Suppose that the proton moves in direction A. The electron must, consequently move in direction B, assuming that the line between A and B is linear.

Moreover, as the proton heads in a direction which is opposing that of the electron, they do so parallel to one another. Meanwhile, the neutron (shell) is moving in a direction perpendicular to the action of the proton and electron...mainly the proton (as it is so much larger than the electron).

This gives us a very precise picture of the basic components of the atom. We have protons which are the essential building blocks of matter. We have electrons which are the very essence of energy. And we have neutrons which are the fluid of subatomic space.

I have one parting thought for you as I conclude this chapter. As the neutron, or weak-force space, moves around the outside of the proton, which direction is it traveling? Therein lies the rub!

INRE: THERMODYNAMICS

There are three acknowledged states of matter which most people can name. I am, naturally, referring to the states known as liquid, gas, and solids.

Each of the three states of matter has the same base element. For instance, in the case of water, we can have ice, water, and steam. All three are water. The difference is in the degree of 'excitation' each represents.

Condensed down to its barest essence, we can say that each state of matter is induced by a thermal variant. Ice is cold. And steam is hot. Or, if you are a real nutcase, we could say that ice is hot and steam is cold.

Actually, there seems to be a fine line between hot and cold. In the most extreme cases, hot can imitate cold; and vice-versa. Yet, the fact remains…temperature plays an integral part in how all matter behaves.

But what do we really know about the concept of hot and cold? We say that cold is really the absence of heat. We use refrigerants to remove heat from refrigerators and freezers. And that all seems simple enough. But what is hot and cold?

When I first discovered that there was such a thing as planetary pairing, I considered that what was occurring at the cosmic level had to be occurring at the sub-atomic level, as well. In fact, this was the inducement for declaring that atoms behaved according to a Dual Mass theorem. It was all so tidy, logical, and counter to what we had been taught.

So it was only logical that I continue with my line of reasoning and include a treatise on Thermodynamic Coupling. By this, I opt not to say that atoms pair up but, to be more precise, they pair up in a purely logical way.

And I decided that I needed to narrow that search down so that I might present a fathomable argument for my thesis. Since I had not, as yet, tackled the problem of the relationship of electricity to magnetism, that seemed the perfect starting point for Thermodynamic Coupling.

Let us say that electrons are hot. We are (overly) fond of saying that electricity is a flow of electrons. And we know that this flow of electrons exudes a specific quantity of heat. So I see little problem with revamping our thinking so that we can accept that electricity is hot.

My next step is to say that Protons are cold.

If scientists are correct, absolute zero is a state or condition wherein all motion ceases. The question is, if this exists, is it in the form of energy, or is it in the form of matter? Or is it something altogether different?

Some scientists will argue that such a state would manifest as anti-matter. Others will speculate that such a state would exist as anti-energy. Both are hypothetical and unprovable.

What is provable is the characteristics of such a state. Such an entity would have to dispense heat faster than the ambient can provide it, or it would have to absorb heat faster than the ambient can provide it. In theory, neither condition can exist.

To expel energy would require energy. And, of course, the acquisition of energy would be self-defeating. Scientists attempt to circumvent these requirements by surrounding their target with material or devices which they hope will draw out all heat (energy). The drawback to this scheme is that it would take something colder than absolute zero in order to achieve that quest.

Am I saying that absolute zero does not exist? Certainly not. I am merely saying that man will never, in our lifetime, achieve it. How can you build something that you don't understand? To me, it is tantamount to believing that a caveman constructed an atomic bomb which obliterated the dinosaurs. Plausible in science fiction, but not tenable in science fact.

Truth is, I firmly believe in the existence of absolute zero. It is an unknown substance which permeates true space. It has the enviable property of instantly transferring heat/energy without absorbing any. It is this thermal property which allows starlight to traverse thousands of light-years from one galaxy to another. Instantly.

People have a really hard time comprehending the idea that light can travel at speeds vastly exceeding 186,000 miles per second. They want to know how such a thing is possible. In my mind, such a feat is elementary.

In absolute zero, scientists claim that all action is stopped. What they mean to say is that nothing, including energy, is moving. All motion is the result of the net force of many forces. Everything, at any given moment, has numerous forces acting upon it.

Though something is moving in a certain direction, does not mean that it is moving because there is a force pushing it in that direction. What it means is that the force which is propelling the object in that direction is greater than the sum of all the other forces acting against that object.

One simple question clears the air as to how light can attain speeds beyond human comprehension. If motion is the result of the sum of all forces acting upon a body, and we eliminate all forces except one, what would you expect to happen?

Absolute zero "freezes" all forces within its specific theater. If something is moving fast enough, in this case light, it escapes the effects of absolute zero. Effectively, when light encounters regions of absolute zero, there is only one force acting on it.

Reverting back to Newtonian physics, Isaac stated that whatever an object is doing, it will continue to do that until acted upon by an outside force. Light is accelerating away from its source. If we eliminate all other forces, light will continue to accelerate.

Newton also said that for every force, there is an opposite but equal force. This implies that there is a force ahead of light which is pushing light in the opposite direction. Let's use a recognizable example so that this concept is clear in your mind.

As a car travels down the highway, it is hindered, or impeded, in its travel by air pressure building up ahead of the car. This pressure is greatest where the front of the car meets up with the air.

If we measure the air pressure at a point 100 feet ahead of the car, we will notice a substantial difference between that measurement and one made at the point immediately in front of the car. In other words, the force of the event drops off the further we go from the event.

Instead of air pressure, let us now consider the velocity of the air. The velocity of the air, at the point where the front of the car meets it, is the same as the velocity of the car. As we go to points ahead of the car, we find that the velocity of the air is considerably less than the speed of the car.

Now, let us return to the light. As light travels it has a force of resistance much like the wind resistance in our example. At all points ahead of the light, the velocity of the resistance is less than the velocity of the light. This slower velocity allows absolute zero to interact in such a way that it reduces the amount of resistance ahead of the light. And the light accelerates.

Why doesn't absolute zero interfere with light? Actually, it does. Light which is traveling at 186,000 miles per second is not fast enough to break the barrier presented by absolute zero. This causes much of the light in our galaxy to reflect back into the galaxy.

Wait a minute. If light has to travel faster than 186,000 miles per second in order to escape the effects of absolute zero, how can any light escape the galaxy? As I said, elementary.

As light approaches the barrier presented by absolute zero, the forces in front of the light begin to crumble like an accordion. The weaker light, that is, the light that does not have sufficient mass/force to push on, will be reflected. Only the light with a magnitude great enough to forge ahead will pass through the barrier.

But how does light get this extra mass/force? To understand, let's look at light from the beginning of propagation until it encounters true space.

Light is strongest when it is released from the source of the light. As it travels away from the source, it is dissipated by various forces which the light encounters along its journey. Some of this light becomes very weak and is easily interfered with.

Two rays of light approach the barrier of absolute zero. The first is only one mile in length. The other is five miles in length. As the force of resistance in front of each collapses, it increases the force immediately in front of each ray. The first ray is turned away because the total force behind it cannot overcome the total force in front of it.

The longer ray hesitates for a millionth of a second while the energy behind it gathered to overcome the energy in front. In another millionth of a second, absolute zero cancels the forces in front of the approaching ray and, by the time the ray reaches the outer edge of the barrier, it has an exponential increase in velocity which allows it to pass through the entire region of absolute zero.

Because of this property of true space (absolute zero), it acts as a barrier to slow moving (sub-light) matter. Given enough time, every galaxy in the universe will attain a state of equivalence or equilibrium. And, thanks to a lack of attraction, all matter will continue to be pushed from dense regions into more sparse regions.

In the meantime, we must reassess our postulate that the speed of light is constant. Whether you examine it in the classical sense, or you subscribe quantum values to it, light and heat maintain an interchange. In fact, you could say that light and heat or one and the same entity.

What happens when heat is generated? Opposites attract (sic). The heat molecule (sic) goes to a place where there is no heat. Let us look at that at an atomic level.

We have two atoms. One has a heat particle and the other does not. Naturally, the heat atom migrates to the atom that does not have one. So far, so good?

Now put that concept in series. This heat particle goes from one to another, to another. It is blazing a trail which ensuing heat particles follow. This is because they all follow the path of least resistance.

Next, we have to deal with residue. Every time there is an exchange of heat, there is a residue which warms the ambient. Since heat is the cause of the propagation of light, every loss of heat represents a reduction in the speed of the light. And this scenario is very accurate within a closed system with set parameters.

So what happens in absolute zero? Two things. In the first scenario, light has decelerated and the interaction between absolute zero and heat particles causes the light to be reflected back into our galaxy.

In the second scenario, has accelerated to a speed beyond 186,000mps. Absolute zero strips away the few remaining heat particles and the light is able to accelerate to near infinity. There is no longer any resistance to hinder, or impede, its motion. And

Newton says that whatever a body is doing, it will continue to do for so long as no outside force (in this case, heat) acts upon it.

So how do I explain this apparent contradiction? What is the nature of this seeming paradox?

What paradox? What contradiction? Are you referring to the whole propulsion thing? Are you insinuating that I am daft because I say that heat is the reason for the propagation of light and, yet, I state that heat is the reason light cannot travel any faster? Is that bothering you?

A rather simple analogy comes from the space shuttle. We use boosters (heat) which enable us to launch the thing into space. At some point during the course of that launch, the boosters empty out and are discarded. What would happen if we did not discard them?

By discarding the boosters, we allow the shuttle to continue to accelerate. If we left them on the shuttle, the additional mass would slow the shuttle down. And there would be a strong possibility that the shuttle could not complete its mission.

There is a threshold at which the presence of the heat ceases to propel the light. In fact, the heat acts as an inhibitor which regulates the speed at which the light can now propagate.

How heat dissipates is very much at the heart of motion as we know it. Thus, it is imperative that we learn the laws regarding thermodynamics. *All* the laws!

ENERGY vs MATTER

There is a great misconception which is clouding modern thinking. This misconception is tactically, and functionally, erroneous, and causes many an educated man to chase off down paths where no clear result can ever be attained. This misconception, of course, is simply that energy can be converted into matter, and vice versa.

Under the doctrine of conservation, that is to say the laws which state that both energy and matter must be conserved, we have come to accept the belief that neither can be destroyed. And this leads many of us to assume that we can convert one into the other.

Energy is energy. Matter is matter. And never the twain shall meet. Not entirely true. You see, each is dependent upon the other. In a sense, one cannot exist without the other.

Matter-antimatter? In a manner of speaking, yes. The reality is that energy's job is to try to convert matter into energy. And matter's job is to try to convert energy into matter. This battle is as old as time, itself, and is inbred into every facet of our lives.

Does not the parent try to make the child unto his own image? Do we not try, on some level, to make friends and family like ourselves? And, if we fail in that task, do we not seek out others like us?

The plain old truth of it is that we use matter to manipulate energy, and we use energy to manipulate matter. We have learned to be adept at work, but we still fail to understand it. Man is truly in his infancy.

The sun spews forth its elixir and we drink it up. We created solar panels, and other such things, so that we can store the sun's energy for later consumption. So why do we not understand that all things do likewise?

Trees store energy which keeps them from freezing in winter. We burn logs to retrieve, or set free, that stored energy. Nothing mysterious there. So why is it that we have such a difficult time applying that logic to everything around us?

There is a relationship of energy to matter which is merely a ratio of one to the other. If we have a certain quantity of matter, then there must be a certain quantity of energy. I suspect that this relationship is 2π. Or, for simplicity, let us suppose that this ratio is six units of energy to every unit of matter.

Space, for all intents and purposes, is a pathway. Just what this pathway consists of has been a mystery. I reckon we shall just call it a point somewhere between energy and matter.

I hate to use the word 'dimension' too freely, but it is the best way to understand space. If we assign three dimensions to matter, and we assign two dimensions to energy, then space must be one dimensional. And it becomes a less complex task to see how all these things interact.

Because there is such a fine line of distinction between dimensions, the established rule seems to be that a dimension can only interact with a dimension that is nearest it. What I mean to say is that three dimensional objects can only interact with either four dimensional, or two dimensional, objects. Two dimensional interacts with either three dimensional, or one dimensional, objects.

This explanation is, of course, solely for the entertainment of those who believe in dimensions of existence. Nonetheless, it makes for some interesting brain fodder. To understand the concept of dimensions, we must first examine the reasons man invented them in the first place.

Nobody knows who the first guy was that came up with the belief in dimensions. Quite possibly it was Euclid as he walked along the X axis and was staring up along the Y axis. Height, width, and depth must belong to some secret society which allows them to exist. Otherwise, why would the Gods allow them?

I am just having fun with you. But I do have my moments where I think those guys smoke way too much. Dimensions. String theory. Where does it all end?

I do, by the way, believe in the idea of string theory. Light rays could be viewed as strings, some parts of space do appear to be stretched out so thin that they might appear as strings, etc., and so forth.

Dimensions, on the other hand, are more fantasy than fact. Dimensions are the tools of the people who lack the ability to formulate any other explanation for the things around them. Take, for instance, blackholes.

More books, movies, and the like, have been made about blackholes than have been made about any other part of science. Blackholes, they tout, are really portals to other dimensions, worm-holes to other galaxies, etc. How can so much be said about something scientists cannot explain?

Blackholes are closest to being described by the worm-holes to other galaxies theory. Remember the theory I stated about our galaxy being surrounded by true (interstellar) space?

The space within our galaxy is restricted. The space outside our galaxy is not. A barrier exists around our galaxy which keeps everything functioning at a slower speed, and lower magnitude, than what we would encounter in interstellar space.

Dark, or black, matter is nothing more than a place where the barrier which separates us from interstellar space is deteriorating. Light accelerates and matter crumbles.

Have you ever observed a dust cloud as it clears? Have you ever watched as the fog drifted away and you could see vast distances again? Dark matter and blackholes are just places where the fog is drifting away.

We are all aware that matter can imitate energy. And we all know that energy can imitate matter. But have you ever given much thought to the role of weak-force space molecules?

Weak force space molecules must be both tactile and elastic; perhaps possessing a high rate of resiliency. It must also get more rigid when compressed. By this, I mean that it gets compacted whenever two bodies collide and stretched taut whenever it encircles large bodies.

I think one of the things that baffles people is the way in which weak force space functions. Although it does tend to permeate matter, it is most effective surrounding the host. Take, for instance, the magnet.

We have this wicked tendency to think of a magnet as a piece of iron through which magnetic force flows. The reality is that the force is outside the iron (as opposed to electricity where the flow is internal). Let me try to paint this picture for you.

When I described an atom, that is to say, when I stated that a neutron was an emulsifier which held the proton in suspension, I might well have said that the weak force space molecules do the same thing to the electrons.

I do not wish to get into a lengthy discussion about this aspect of the atom at this particular time. This is partly because the theory is a fairly recent discovery and I have not worked out all of the bugs. Let's suffice it to say that the electrons which flow inside of a piece of wire are "uninsulated" and the ones flowing on the outside of the wire are insulated.

If weak force space wants to wrap itself around electrons, then this force would track the electrons within the wire. As soon as the electrons were discharged (ostensibly into the air) weak force space would rush to envelope them. Or something along those lines.

Just one more thing to think about.

THE VARIABLE SPEED OF LIGHT

I am not the first person to suspect that the speed of light is not constant. I know of one person who suspects, or suspected, that the speed of light was, at one time, much greater than 186,000 miles per second. He attracted a lot of attention with that one.

But the man was wrong. The speed of light was not faster than 186,000mps. It still is!

In the preceding chapter, I explained how the speed of light can be advanced. In this chapter, I wish to expound upon how light is retarded.

We live in a galaxy that is spinning. Every inch of space within this galaxy is curved. That is a direct consequence of the motion of the galaxy. It is also the reason why light travels in a curved motion (arc).

The relationship between energy, matter, and space, have become so calculable that we need only know the distance across a galaxy, the distance between the arms of the spiraling galaxy, and the angle of those arms (relative to the core), in order to give the speed at which the galaxy is rotating on its axis.

None of this means much to people interested in knowing the nature, more specifically speed, of light. To those of us who understand how the entire system works, the realization that light must travel under the most adverse of conditions (closed system, curved space, pollution) assures us that light has been slowed down.

Scientific researchers have been dogged for more than a century by the concept of relative space wherein the speed of light is constant. Likewise, these Theoreticians have been mystified by phenomenon such as the aberration of starlight. We can, at last, alleviate the apprehension brought on by those perturbations.

For many years, I stared at the night sky in sheer disbelief in what I was seeing. I wondered, as had many before me, how it was even possible for starlight to reach the Earth. That light, came from such a vast distance that it should have dissipated or, at the very least, been disrupted or interfered with long before completing its journey.

Then one night, I watched intently as a manmade satellite approached a not-particularly-bright star. For some inexplicable reason, I speculated about what would happen when the two objects converged. While ruminating, the event passed, and that created considerable turmoil over what had been seen. Instead of blocking out the light from the distant star, the craft seemed to veer out around it in a rather wide arc.

The light from that star was clearly "bundled." Yet, it defied logic. How could light, which originated so far away, retain a magnitude sufficient to "push" aside a solid object?

For me, there was only one viable conclusion which would satisfy each of my questions. The solution was that light from those stars was traveling at speeds far greater than those observed in classical physics. But how could that be?

At first, I speculated that light was one-dimensional. As long as it could travel in a straight line, its speed would approach infinity. However, in a closed system, such as the Milky Way, it was forced to traverse along a curved path through space. Aha!

Even if we ignored the simple math of angular momentum, or that of inertia through mixed mediums, we could not ignore the flirtations with other dimensions. Light travels slower in curved space because it is being forced into an additional dimension. And a given unit travels considerably slower in two (or more) dimensions than it does in one.

Some of the most profound discoveries are those which were chance encounters. Put another way, if you went out in search of gold (ore), and you did not know anything about gold, how would you know whether, or not, you had succeeded?

The answer, of course, is persistence. When everything else fails, try again. Always, try again.

The light was behaving as an automobile would under similar circumstances. As long as it was traveling in a straight line, it continues to accelerate. Within a curve, the car will slow down. Nothing so mysterious about that. But, for some inordinate reason, theorists prefer to ignore the obvious.

As light prepares to leave this galaxy, it encounters resistance from finite particles within the galaxy. At the walls where the galaxy encounters interstellar space, thermodynamics kicks in an those particles are stripped away.

As the resistance which pushes against the light is stripped away (ostensibly by freezing it in absolute zero), light begins to accelerate. By the time the light escapes this galaxy, it has already accelerated beyond the accepted maximum of 186,000mps.

And light will continue to accelerate until it, once again, enters another galaxy, or closed system. I would venture to say that it takes less time for light to travel between galaxies than it does for light to travel from the sun, or source, to the galaxy edge.

But then you knew all of that already; didn't you?

THE ASTEROID BELT

There has been much speculation about what happened which caused the creation of the asteroid belt. Most people are of the opinion that the asteroids exist because of some collision between two, or more, celestial bodies. Perhaps. But not likely.

I have two theories which fit the bill. Unfortunately, I do not have sufficient data with which to select one over the other. Therefore, I will leave the decision making processes to those in better positions with which to make the choice.

First, let us consider the available data. On one side of the asteroid belt, we have Mercury, Venus, Earth, and Mars. All are under 13,000 km in diameter. Compare that to the outer four; the smallest of which is close to 50,000 km. What can we deduce from that information?

The solar system has many safeguards built in so as to ensure that the system remains around for many, many, years to come. This fact, more than any other, leads me to believe that the asteroid belt plays a pretty major role in this process of preservation. And this realization pushes us towards two possible scenarios.

The first scenario is that an invisible barrier exists which prevents any object above 13,000 km from entering the inner sanctum of the solar system. This device would keep the sun from being assailed by an object big enough to destroy it. Boy, I hope Roche doesn't get credit for that one.

I have little doubt, actually no doubt whatsoever, that the asteroid belt is the physical remnants of a cataclysmic event caused by an invisible barrier which crushes large celestial bodies into much smaller ones. That much of the theory is intact. The problem I am having difficulty with is in determining the exact function of the barrier. More than likely, there are two functions and not two theories.

So we come to the second (part of the) theory. The Great Barrier Wall (as I prefer to call it), breaks large objects into smaller objects for fuel. This means that, as the sun consumes itself and loses mass, it needs more fuel. This fuel, the Barrier supplies in "bite-sized" pieces which will not greatly affect the sun's performance.

Two functions of survival. Which is correct? Probably both. If so, we are slated for dinner. How humbling is that?

But you need not worry that much about it. I suspect that such conjecture is purely science fiction. And, if it turns out to be science fact, I think that the sun will only consume "dead" planets/bodies in space. Under that postulate, Mercury, our moon, and, eventually, Venus...but not the Earth.

Why Venus? Why not? Venus was once a thriving metropolis; much like the Earth. Water, vegetation, various forms of animal life. Perhaps, it once held a civilization comparable to Earth's.

Legends abound in which advanced civilizations are alleged to have existed on Earth. And as many legends exist which purport interplanetary space travels. It is not such a stretch of the imagination to envision those things. After all, we could, very easily, do it today...if we had someplace to go.

If Venus was in the region of space currently occupied by Mars, and some catastrophic event forced it out of its orbital path, it is possible that the event wiped out life (as we know it). And any survivors from that event would have perished long ago under the current conditions.

Here is another interesting tidbit: what would happen if the object that struck Venus were to go into the sun? Let's look at history.

History tells us that an advanced civilization known as Atlantis used to exist many thousands of years ago. Assuming that this was a colony that came here from Venus (while Venus was in the Martian orbital path), it would be a safe bet that they were advanced enough to see a huge monolith bearing down on their home planet.

Now, knowing that a huge asteroid was about to strike their planet, they could have deduced the possible outcome(s) of that event. And, having made so many predictions, they would have come to the conclusion that the Earth would not escape unscathed.

Perhaps a large chunk struck the Earth. We do have a wobble, and such an event would certainly account for that. And we do know that something would have struck the moon to stop it from rotating on its axis. So the possibility exists.

My personal belief is that a large chunk, quite probably the offending protagonist, sailed into the sun. It was not big enough to do the sun any harm. However, it was large enough that it caused a sustained solar flare-up which did cause the Earth numerous problems.

What would happen in the event of a sustained solar flare-up of the magnitude of the aforesaid event? Well, the Earth's temperature would greatly increase. Wide-spread flooding would be expected. In fact, it would seem that this flooding was predicted and many people on our planet attempted to prepare for that flood.

Is the Biblical flood the self-same flood created by the event of which I speak? Again, I must reiterate, it is a pretty safe bet.

Assuming that the Biblical flood is the same event that caused the relocation of Venus and its moon, Mercury, then we have a pretty precise date for that occurrence. And we can evaluate the damages well enough that we might be able to brace ourselves for the next event.

One important thing to remember about the Great Barrier (asteroid belt) is that it takes large bodies and crumbles them into smaller bodies so as to protect the sun. That does _not_ mean that we are protected by the barrier. What may be small, relative to the sun, is definitely not small relative to our fragile planet.

GRAVITY, FORMULAS,& PREDICT...

You know, I almost hate writing this chapter. I have advanced some of the most ingenious theories to come down the pike in a long while. But scientists have become like half-breeds. One part is vulture and the other part is lawyer. Either way, the outcome is the same.

I advance the theories. If I fail to list even one prediction, some wise guy will come along and make a prediction for me and then claim that my theory was really his theory. That, as they say, is all hogwash.

Many theories have come along without predictions being directly attached to them. After all, it takes a lot of thought just to get the theory down. But none of them were robbed. Of course, they were honorable men who were in the midst of other honorable men. Somehow, I do not feel that same kinship.

Times have changed. And so I must attempt to make some predictions which my theories may substantiate. Please note that this is only a partial list as I do not intend to waste a whole lot of time making predictions about the obvious.

My first prediction is, naturally, that Venus has signs on its surface of a past collision with a large planetoid of some sort.

I also believe that Venus will reveal to us that there used to be an advanced civilization on the planet. How exciting is that?

Sometime around 2014, the first man-made object (spacecraft) is slated to leave our galaxy. I predict that it cannot exit. It will either be crushed, ripped apart, or just disappear.

My theory about the emulsion of the proton inside of the neutron will lead to a new theory which explains the nature of radioactive material. At least some part of that theory is going to deal with the manner in which radioactive material is produced. Let me clarify this with my own theory.

Let us suppose that the Earth was in the region of space currently occupied by Mars. Venus would, then, be sitting in the region in which the Earth currently resides.

As you would expect, Venus was the habitable planet and did, in fact, have an advanced civilization. The Earth, at that time, was probably not inhabitable (as in current day conditions).

Now, the planetoid crashes into the Earth, pushing it towards Venus. Under my theory of repulsion (as opposed to gravitational attraction), the Earth need not collide with Venus in order to push it out of its orbit.

This scenario satisfies many unknowns. We can pinpoint the region of contact between our planet and the invader. How? Simple. I believe that radioactive material is just material that was rapidly compacted under tremendous pressure in such a way that the balance between energy, matter, and space became imbalanced. To find the point of impact, we need look no further than the largest concentrations of radioactive material.

By my estimation, the largest concentration of radioactive material is in north America. To be specific, it is in Canada. And that fits, perfectly, with the wobble scenario I touched upon.

This scenario gives support to my planetary pairing theory. Up to this point, the pairing has been shield by the disparity of the inner four planets. By eliminating Mars and Mercury, and moving Earth and Venus back into their original positions, we put everything back on track.

Remembering Newton's theory that whatever a thing is doing, it will continue to do until acted on by an external force, we can easily see that the space around the Earth would continue to rotate at a period of approximately 24 hours per current Earth day. When the Earth was bumped out of orbit, its moon, Mars, would have gotten locked into its current position. And, as a direct consequence of the rotating space, begin to rotate at 24 hours per day.

I predict that Mars will slow down in its period of rotation. So? As it slows down, it will develop a stutter step like that of Mercury. Eventually, it will drift out of orbit and move towards the sun.

I suppose that I had better explain why Mars will slow down and drift out of orbit. Earlier in this book, I stated that an electron will periodically strike the surface of a proton (nucleus of the atom). What I did not tell you was that it occurs most frequently in metal (because it is the densest material that can remain liquid…among other things).

Metal allows the free flow of electrons. In a molten state, this free flow is at optimum and electrons plow into protons at a greater rate. It is this action which causes the planets to rotate.

As you might expect, smaller celestial bodies cool down quicker than their larger companions. The moon and Mercury are prime examples of solid core celestial bodies that have cooled and all but stopped.

For your information, scientists have concluded that the moon is, indeed moving away from us. It is inevitable.

Getting back to Mars, because of its relatively tiny size, Mars should not be rotating as fast as it is. We can, safely, conclude that it has been acted upon by an outside force.

Some men of intelligence will claim that there may be a large deposit of metal near the surface in a discreet location. The interaction of this deposit with the sun's energy could account for some of the motion. However, I will stick to my theory regarding the motion of the space around the planet.

Remember that the weak force space that surrounds a body in space is very weak. Apparently, in a collision between two bodies, the ratio of weak force molecules to matter gets changed. To be exact, when Venus was pushed out of its orbit, it left behind the majority of its weak force space molecules.

The orphaned space molecules attached themselves to the nearest body (Mars). Because the space molecules were rotating at 24 hours per day, Mars was forced into compliance.

The question arises, if the weak force space molecules are the reason for the planetary rotation, why consider the state of the core of the planet at all? Think in terms of a capacitor. Think about how a capacitor is made. Ditto for batteries.

A capacitor is constructed of two, or more, conductors separated by nonconductors. Batteries have alternating layers of similar nature. The basic idea is to get energy flowing, or available to flow. And this is the idea behind planetary motion. Energy must flow.

Obviously, the better energy flows, the faster the motion. Just something to think about.

Getting back to predictions, I predict that my theories will lead to the discovery of a propulsion system that utilizes light as its driver. Light accelerates in absolute zero. The key to such an engine is to get either space or a crystal cooled down. A super-cooled crystal seems like the most logical. Can you envision that? Imagine, a flashlight propelling a car through space!

I predict that scientists will prove that my weak-force space molecules can imitate either energy or matter.

My formula for calculating gravity is so accurate that I predict scientists will be using it and will discover that their calculations for the planet Saturn are off substantially.

Let's take a closer look at the mathematical side of my formula for gravity. How does it compare with the classic formula?

With the exception of the sun, the following facts were taken from a book in the Teach Yourself series, called Planets. The book was written by David Rothery. The book says Mr. Rothery is (or was) a teacher of Geology and Planetary Science at the United Kingdom's Open University. I thank him for both the book and the data.

Planet	Mass	- Eq. Radius	- Density	- Gravity
MERCURY:	.055	2439	5.43	.38
VENUS:	.815	6052	5.20	.90
EARTH:	1.000	6378	5.52	1.00
MOON:	.0123	1738	3.34	.17
MARS:	.107	3396	3.91	.38
JUPITER:	317.70	71492	1.33	2.36
SATURN:	95.16	60268	.69	.92
URANUS:	14.54	25559	1.32	.89
NEPTUNE:	17.14	24766	1.64	1.12
SUN:		700,000	1.41	

In the above chart, the radius is the Equatorial Radius and is measured in kilometers (km). The Density is given in g/cm^3; which I believe is grams per cubic centimeter. And the gravity is the ratio (percentage) of each to planet Earth's gravity.

Radius X Density = **Gravity**
Gravity / Radius = Density

Using the formula stated above, let's solve for gravity. For example, we take the radius of Mercury (2439) and multiply that times Mercury's density (5.43) and we derive at 13,243.77 and we put that in our chart. Now, let's do that for all the planets, our moon, and the sun.

Planet	r times d:
MERCURY:	13,243.77
VENUS:	31,470.40
EARTH:	**35,206.56**
MOON:	5,804.92
MARS:	13,278.36
JUPITER:	95,084.36
SATURN:	41,584.92
URANUS:	33,737.88
NEPTUNE:	40,616.24
SUN:	987,000.

So far, so good. Right? Well, almost. Many of you are now in the dark. You are wondering what does that mean? So let's convert those numbers into something you better understand.

Since we all know a thing, or two, about the gravity of the Earth, we can use that as our measuring stick. Now, in order to derive at a comparison value, we simply divide the planet we wish to compare, by the number of the planet we want to compare to, and we get the relative value of these numbers.

The rd for Mercury is 13,243.77 and the rd for Earth is 35,206.56 and so we must now divide as follows:

R times d Mercury divided by r times d Earth. 13,243.77 divided by 35,206.56. The answer is .376 and is put in the chart as .376 of the Earth's gravity. So Mercury has just a little over one-third of the gravitational strength of the Earth.

Now let us do that for all the planets.

Planet	r times d:	gravity (Ratio)
MERCURY:	13,243.77	.376
VENUS:	31,470.40	.900
EARTH:	**35,206.56**	1.00
MOON:	5,804.92	.165
MARS:	13,278.36	.377
JUPITER:	95,084.36	2.701
SATURN:	41,584.92	1.181
URANUS:	33,737.88	.958
NEPTUNE:	40,616.24	1.153
SUN:	987,000.	28.03

Next, let us take those percentages and put them in next to their calculations. Theirs are the numbers on the left under the header "Gravity." Mine are on the right, in parenthesis.

Planet	Mass	Eq. Radius	Density	Gravity
MERCURY:	.055	2439	5.43	.38 (.376)
VENUS:	.815	6052	5.20	.90 (.893)
EARTH:	1.000	6378	5.52	1.00 (1.00)
MOON:	.0123	1738	3.34	.17 (.165)
MARS:	.107	3396	3.91	.38 (.377)
JUPITER:	317.70	71492	1.33	2.36 (2.701)
SATURN:	95.16	60268	.69	.92 (1.181)
URANUS:	14.54	25559	1.32	.89 (.958)
NEPTUNE:	17.14	24766	1.64	1.12 (1.153)
SUN:		700,000	1.41	(28.03)

If you wanted to convert those percentages, or ratios, to numbers you are more familiar with, just multiply by the earth's gravity. For example, if you wanted to know Mercury's precise gravity in m/s^2, multiply Mercury's ratio to earth (.376) times 9.78 (m/s2), and you get: .376 x 9.78 = 3.677m/s^2.

Many of you have, undoubtedly, noticed the discrepancies between *their* classical calculation of gravity and my, more precise, calculations. Theirs only works on some things. And mine works well for all things.

I next want to focus on certain patterns which presented themselves when I rearranged the planets data in order from smallest radius to largest radius.

Planet	Mass	Eq. Radius	Density	Gravity
MOON:	.0123	1738	3.34	.17 (.165)
MERCURY:	.055	2439	5.43	.38 (.376)
MARS:	.107	3396	3.91	.38 (.377)
VENUS:	.815	6052	5.20	.90 (.893)
EARTH:	1.000	6378	5.52	1.00 (1.00)
NEPTUNE:	17.14	24766	1.64	1.12 (1.153)
URANUS:	14.54	25559	1.32	.89 (.958)
SATURN:	95.16	60268	.69	.92 (1.181)
JUPITER:	317.70	71492	1.33	2.36 (2.701)
SUN:		700,000	1.41	(28.03)

Now, let us extract the pertinent information with which I wish to establish a pattern. In other words, let's list the radius from smallest to largest.

Then, next to those numbers, let us place the ratio as compared to the Earth's gravity.

Planet	radius	gravity (times Earth's)
MOON:	1,738.	.165
MERCURY:	2,439.	.376
MARS:	3,396.	.377
VENUS:	6,052.	.893
EARTH:	6,378.	1.00
NEPTUNE:	24,766.	1.153
URANUS:	25,559.	.958
SATURN:	60,268.	1.181
JUPITER:	71,492.	2.701
SUN:	700,000.	28.03

The first thing you notice is that gravity increases with the increase in size…until you get to Uranus. So I must predict that their data for either density or radius, or both, is wrong. If all other data is right, then Uranus' gravity will lie somewhere between 1.15 and 1.18 times the Earth's gravity.

My formula for gravity is not only simpler to use, it is more accurate. However, it only defines the relationship of matter to energy and of matter to space. It has failed, indirectly, to establish the relationship of energy to space. Or has it?

The sun rotates at almost 28 days and has 28 times the gravity of the earth. Does this imply that gravity is cumulative? That is, 1/28 rev of the sun equals $9.8m/s^2$. Compare that to the moon's rate around earth.

Throughout this book, I have offered you clues that will summarize the exact relationships between energy, matter, and space. Let's review some.

I stated that there were two types of force. I stated my belief that there is, possibly, a chance that one form of space is synonymous with light. I mentioned a saturation point. But failed to explain that in any complete, or satisfactory, manner.

Saturation. Consider each planet to be a battery. Each battery is contingent on physical parameters which dictate the amount of "charge" that each can maintain. In other words, each planet is limited and reaches a saturation point. I will leave it at that.

Distance has no effect on the magnitude of the charge. In this context, the charge is gravity. This contention directly opposes the traditional thinking as handed down by Newton. Distance can affect the rate of charge, but not the magnitude.

There, I have provided all the tantalizing clues which will help you derive a proper formula for the relationship of energy to space.

Oh, and did I tell you about the effect of the sun's motion? Tee hee.

I'm not being very nice, now am I? You're right. More clues are in order. Consider the following:

There is a relationship of space to gravity which dictates the amount of matter which each can contain. For instance...

Planet	r times d:	gravity (Ratio)
MERCURY:	13,243.77	.376
VENUS:	31,470.40	.893
EARTH:	**35,206.56**	1.00
MOON:	5,804.92	.165
MARS:	13,278.36	.377
JUPITER:	95,084.36	2.701
SATURN:	41,584.92	1.181
URANUS:	33,737.88	.958
NEPTUNE:	40,616.24	1.153
SUN:	987,000.	28.03

…when you add the r times d calculations for gravity for the bodies within the first five AUs of space, this being the Moon, Mercury, Mars, Earth, Venus, we get a figure of 99,004.25. For the time being, we shall ignore the asteroid belt.

Now look at the next 5 AUs of space. This consists of Jupiter and its moons, plus various others. Jupiter's radius times density is 95,084.36 and that is very close to the 99,004.25 that we had for the other 5 AUs. We can easily see that the discrepancy may be because the asteroid belt belongs to Jupiter, or we can add the moons to the total, or we can search for the real answer.

Jupiter is the second largest body in our solar system. In classical physics, we can safely deduce that the greatest force will occur between the two largest bodies. Hence, the greatest force exists between Jupiter and the sun.

Let's see if we can explain the solar system, gravity, and light, unilaterally and tangentially.

Jupiter is the barrier planet. Beyond Jupiter, the planets get smaller and smaller. However, the gravitational force still exists. By adding Saturn, Uranus, and Neptune, we get a sum of 115,939.04 (not counting the moons).

Let's examine the next table for a more precise explanation of the Solar System and its functionality. This table will enable us to make many predictions for the entire system.

Planet(s):	total gravity:	total radius:
Mercury plus Venus	1.269	8,491. km
Earth plus Moon	1.165	8,116. km
Earth plus Mars	1.377	9,774. km
Merc + Ven + Moon	1.434	10,229. km
Total of all inner	2.811	20,003 km
Total all inner + Jup	5.512	115,087.36 km
Total of all outer	3.292	115,939.04 km

The symmetry within the solar system is as intricate as the movement within a watch. We have no precise methodology available to us by which to be exact in our measurements of planetary density. Nonetheless, we can calculate to a high degree of certainty.

Under the premise of Planetary Pairing, we know that there existed six planets in our not-so-distant past. Venus was paired with Earth, Jupiter with

Saturn, and Uranus with Neptune. But what about the moon, Mercury and Mars?

Gravity for Earth equals 1. Gravity for Venus equals .893. However, when we add the gravity from our moon to Venus, the combined gravity is 1.058 or, in relative terms, equal to that of the Earth. In other words, the Moon was orbiting Venus so that Venus was perfectly matched with the Earth.

At that time, Venus and our moon were occupying the same space where we are now. The Earth, was where Mars is. Because the Earth and Venus were so closely matched, both were rotating on their axis at about 24 hours each. Then came disaster.

A large celestial body, presumably Mercury, strikes the Earth a glancing blow which pushes the Earth towards Venus. The gravitational force exerted by the Mercury/Earth pair, causes Venus to be pushed away from its orbit and the Earth replaces it.

The moon then collides with Mercury and both stop rotating on their respective axis. Mercury strikes Venus on their way to their present positions, and neither of them rotates very fast.

Meanwhile, back at the orbital path that the Earth used to occupy, Mars is drawn in from Jupiter, or the asteroid belt, and rotates at the former occupant's speed. Hence, twenty-four hour days.

Now, what has happened is Mercury is paired with Venus as Planet and Moon, Whereas the

Earth is paired with Mars as the secondary planet and moon in the Planetary Pair. Huh?

Mercury's gravity is .376, relative to Earth. Venus' gravity is .893 (again, relative to Earth's). That is a total of 1.269 times the Earth's gravity. Earth is 1 (relative to itself), and we add that to Mars' gravity, .377 (relative to Earth. That is a total, for the Earth/Mars set, of 1.377.

We next compare those two values. 1.269 is only .108 less than 1.377. Extremely close. And extremely important!

You will have noticed that we have , all but, ignored the presence of our moon. I did that omission deliberately so as to stress its importance. Rather, I need to explain its orbit around the Earth.

Consider this, when the moon is on the side of the Earth nearest to Venus, we add its gravitational force to the Venus/Mercury pair. That is .165 (moon) added to 1.269. Total gravitational force for that triad is 1.434. Compare that to the Earth/Mars pair, 1.377, and we begin to see the picture.

You still don't get it? Okay, let me ask you a question. We now know that gravity is a force of repulsion. That means one body pushes against another. We have two forces. One force is a magnitude of 1.434 and the other has a magnitude of 1.377. Which way is the moon going to go while it is in between these two forces?

Obviously, the moon is going to take the path of least resistance. But what happens when it gets to the side nearest Mars?

Yep, you guessed it. The Earth and Mars set has the additional force of the moon. Now the forces are 1.377 plus .165 is 1.542. So the pressure is on the moon to go to the other side of the Earth.

From this fact, we can deduce several things. One, the gravitational force is accumulative for space. Two, it increases in the direction of motion; provided that motion is in the direction of a source of gravity.

Some of you have noticed that the moon is in between Mars and the Earth at the time I made the last calculation. You presuppose that I have made an error. Indeed, I was almost convinced of that myself. However, when you think of it as a fluid chain, and the moon as its weakest link, then you start to see the picture.

Questions arise about the influence (or lack of influence) of both the sun and Jupiter. Do not forget that this force is tangential to the point of application. In addition, the sun's influence is all but nullified by the radiant energy reflected back in from the outer limits of the solar system.

There are other things we can glean from this information, such as latent energy effects, but I will save that for another time; another book. Right now, I just want to point out a few more tidbits.

In my table of facts, you will note that the combined gravity for Mercury, Venus, and our moon, is 1.434, and the total radius for these three entities is 10,229km. Those totals are almost exactly half the value of the totals for all the inner planets. Total gravity is 2.811 and total radius is 20,003km. The disparity is neglible.

The total gravity for the inner planets comes to 2.811 and that almost matches Jupiter's gravity of 2.701 times the Earth's. That isn't an accident.

Look at the totals for Jupiter and the inner planets. Gravity is 5.512 and radius is 115,087.36km. Compare that to the totals for the other three giant planets. 3.392 total gravity and 115,939.04km total radius.

So? The similarity in radius is exactly as I would expect to find it. We live in a balanced system. But something is wrong.

No, I am not referring to the difference in radius verses gravity. Let me clarify that. The radius of the inner planets totals only 20,000km. Yet it has as much gravitational force as Jupiter, and Jupiter is 95,000km. That is easily solved.

Jupiter is a ball of gas that is less dense than the inner planets. When, and if, it cools down, it will probably shrink to a ball around 20,000km in radius.

The problem that disturbs me the most is that the total gravity for the outer planets is 3.292. Some of you may think it too low. Others will think it is just right (considering that the three outer planets in this calculation are all gas balls). Personally, I think the figure may be too high.

If Jupiter is the dividing point between inner and outer, there should be a closer balance between the gravity of the inner verses the outer. Perhaps there is. You see, there are times when Uranus, Neptune, etc., cut across the path that Jupiter takes around the sun. And, at those moments, gravity inside is greater than gravity outside.

Another explanation that I am sure people will embrace is the fact that the outer planets are dispersed across so much more space that the gravity does not concentrate as it would inside the barrier. Possible, but not plausible.

I believe that space is stretched taut enough that any gravity is dispersed across the breadth of space. This opposes the notion that gravity is like a ray connecting bodies.

But enough of that. I do not wish to divulge too much at one time.

In my theory regarding the relationship of energy to space, I will conclude that the energy within the Jupiter barrier is equal to the energy beyond the Jupiter barrier. This means that the asteroid belt belongs to the inner 5 AUs. It also means that Jupiter takes advantage of both the asteroid belt and the outer moons/planets, to offset the two.

In my next book, I will break this down even further and clarify the whole shebang. I just thought that you might enjoy the opportunity to match wits (I.E.- solve the little riddle).

The reality is that Jupiter and its moons were once "paired" up with the other three "giant" planets. In this scenario, Mercury, Venus, Earth, and Mars, were all, originally, moons of Jupiter. If you desire a very precise picture, mathematically, of the early stages of our solar system, this is where you need to start.

I could go on with the predictions. I will even attempt to do so in a future book. Trouble is, I am a regular stiff working long hours at a menial job which does not allow me time to write. But they cannot impede my thought processes.

I send this book off to be published. I realize that it is incomplete and lacks in organized structure. My apologies.

The things I advance in this book are important; at least to me. If they are as important to the scientific community, then I have done a marvelous thing in releasing before completion. If not, then it makes little difference.

I have tried not to occlude anything of any significance. However, when a man finds himself hard-pressed for time, omissions may have been made. Again, I apologize.

I thank you for having read this much and pray that you glean something from it. The rest of this book was my early work and is insignificant...other than to reveal the path I took to get from where I was, to where I am. I include it for posterity.

Please do not think it vanity that I consider my work to be so important. A glimmer of vanity is at work, true enough, but that was never my primary motivation.

I once read that scientists, and other researchers, were sorely disappointed that little of Einstein's early works survived. Little is known of those processes and excursions into parallel ideas, theorems, etc. I did not wish to repeat that error of omission.

On top of all that, if I were truly a vain person, I would have selected my best work (I.E.- the work that I feel is right) and excluded any that I knew was erroneous. But I know that would be a cardinal sin as the wrong paths are as important as the right ones. We learn from those.

And wouldn't it be ironic if one of my earlier thoughts proved to be the most fruitful…while I, myself, selected the wrong one!

\

PART

TWO

Old School Thinking (my early
thoughts)

This part of the book originally appeared in my book, 'Unified Field Theory.' I am including it here as a hint of the evolution of thought. It is for entertainment value more than educational value. Even so, it may just be interesting enough to justify the read.

INTRODUCTION REVISITED

It has been said that any idea which can be proven is, thusly, elevated to a position in science. Conversely, any idea which, largely, remains unproven, gets a minor notation in the world of philosophy. As a direct consequence of this method of classification (segregation), God is placed on the back burner while the lofty atom assumes a primary role in creation.

Not many people, these days, has ever made that comparison. God assumes a position in each of our respective lives based upon a purely arbitrary inclination to accept, or reject, the concept of an Almighty Creator. Atoms, on the other hand, are much touted. And we would be hard-pressed to find even one person (in the so-called civilized world) who would dare to refute the very existence of an atom.

Personally, I find it a strange irony that God has been pushed out of our schools while, simultaneously, the atom enjoys unabated universal acceptance in these self-same institutions. The irony? We can neither see God, nor the atom. Both, logically, belong in the realm of philosophy as we cannot prove the existence of either.

I am not writing this book to prove, or disprove, the existence of God. Nor am I writing this book to prove, or disprove, the existence of the atom. What I *am* striving to do is to get man to do something which he seems to have forgotten how to do. I am, of course, referring to a little used concept called 'thinking.'

So why has mankind, as a whole, forgotten how to think? Well, the short answer is because we have all been brainwashed or, otherwise, duped, into accepting hypotheticals as facts...the existence of atoms being only one of these facts.

I believe that I have given the answers to a great many of the questions which have perplexed man ever since the advent of science. However, what I believe, and what I can prove, may or may not be enough to revolutionize the field of science. Be that as it may, I am assured of one thing. Actually, two things.

First, and foremost, I have provoked you into using your gray matter. You will no longer be content just to accept things as your peers tell you to believe.

Second, I have steered you into the right direction for making new discoveries. I have done this by revealing what I think is the true nature of the universe. And, as I advanced each theory, I endeavored to reveal its significance, how it might impact some other field, and how we might prove its validity.

As much as I hate to admit it, I have no (as of this time) fully developed and/or useful description of the atom. I am not entirely in agreement with the Bohr model. In fact, all I can say, at this time, is that I feel an inclination to abolish Mendeleev's Periodic Table as being erroneously ineffective and grossly misaligned.

Wait a minute. Hold the phone. Stop the presses. Something's wrong. How can I claim to have a "Unified Field" theory and, yet, I exclude a depiction of atomic or sub-atomic particles?

Well...whole mule! I never said that I could not, or would not, attempt to construct a model of the atom---assuming that one exists. What I said was that I do not, currently, have a clear idea of how the atom may be constructed. Please allow me to explain this apparent contradiction.

Albert Einstein, among others, felt that there should be a formula, or two, which would explain everything in the known universe. Such formulas should define everything from stellar systems to miniscule Quarks. This, by definition, would be the Unified Field Theory.

By using simple logic, I advance theories which accurately defines gravity, black holes, etc. It therefore follows that such theories ought to define sub-atomic particles. The exception to this hypothesis is just in the possibility that the sub-atomic world may behave entirely different than its cosmic brothers. We shall see.

For many years, I felt I was on the right track. But there was always one thing, in particular, which haunted me. My nemesis was in the realization that, under Newtonian precepts, starlight could not, possibly, travel all the millions of light-years to Earth. Yet, I have witnessed this phenomenon for the majority of my life.

Einstein's nemesis was in trying to define gravity in such a way that it would work for all things, large and small. He could see a cause, and he could feel an effect, but he just couldn't connect the two.

Starlight. Therein lay the riddle of the universe. How could it travel so far, night after night, day after day? I was stumped.

Scientists or, I should say, mathematicians, have busied themselves crunching numbers. You could say, "They cooked the books." Instead of searching for the definitive answer, they provided us with a mathematical fantasy. Starlight, according to them, gets here because stars are really superhot suns.

In a word, bullshit! Starlight gets here because it is really, really, fast---and "True" space is something other than curved space. Yet, even that wasn't the whole solution. A third factor lingered in those dark regions which we often choose to ignore because to accept it would border on the absurd.

Funny. Rod Serling is never around when you need him. Probably off playing some crazy variation of chess with Mr. Hitchcock.

SPACE AS A 3-D ENTITY

Even though I developed my "Push" theory a long time ago, and it solves a great deal of problems, I encountered several more which were not resolved by Push alone. Amongst these problems were the manner in which distant bodies (in space) seemingly acted upon one another, and, also, the way in which magnets seem to be polarized.

The problem of magnetic attraction could be explained by a combination of field and spin. The larger problem of cosmic attraction relied, almost entirely, on field. In either event, the importance of space could not be ignored. Einstein's declaration that the Ether was of no consequence, as far as Relativity was concerned, was a monumental blunder.

Incidentally, it is historical fact that Einstein was opposed to the idea of quantum mechanics. It is also historical fact that Max Planck introduced the concept of quanta five years before Einstein introduced Relativity. How Einstein ever got credit for 'fathering' quantum physics eludes me (not).

Be that as it may, space has become much more than a coefficient, or measure, of time. By my estimation, spatial function(s), or qualities, are much more important than (re)constructing the atom. Indeed, space may well be the nucleus of the atom.

There is a tendency for us to look at space and think of it as being one dimensional (two, if you count time). We have exhibited so much contempt for space that we readily dismiss it as meaningless or insignificant. It is time to go back to the beginning.

Picture, if you will, an immense ball of matter. I mean unfathomly huge. This massive body, according to those who support the Big Bang, contained everything within the universe. Do you know what it did *not* contain?

If you guessed 'space,' you are absolutely correct. Space simply did not exist.

Now, ignite the Big Bang. See how everything flew apart? Big chunks of this. Small chunks of that. And, eventually, very large patches of space.

Are you following this line of thinking? Before the advent of Big Bang, there was no such thing as space...or, rather, large bodies separated by space. After Big Bang, lots and lots of space. Pretty simple, huh? Not really. Let's jazz it up a bit.

Man has a wicked tendency to do one of two things. One, we try to simplify everything by rationalizing it away. Two, we complicate things by irrational thought. Genius, therefore, becomes a matter of connecting the rational with the irrational.

Many theorists postulate that Big Bang was really two explosions which occurred within micro-microseconds of each other. The reality is less complex.

Ever hear of a supernova? Could that not, rightly, be called a 'Big Bang?' Instead of chasing the absurd, shouldn't we be accepting the obvious? Instead of creating a universe from one huge explosion, which could not happen with infinite mass, with infinite energy, and infinite time, why not enlist the idea of many, many, Big Bangs?

Every galaxy began because of a Big Bang. Farfetched?

Not really. I heard, recently, that astronomers are watching while two galaxies collide. Two galaxies! Wow! Can you imagine our sun ramming into another sun? Then repeat that process for every sun in each galaxy. Truly awesome!

Instead of complicating matters, didn't I just simplify them? Actually, no. You see, I just stuck an infinite number of events inside a much larger event which, in all probability, is encased inside a much larger event, ad infinitum.

Each galaxy is a closed system, and must be treated as such. This creates a singularity which has not played out yet.

The Big Bang of our galaxy has created an invisible wall around the galaxy wherein there exists a pressure on all bodies within the system. This pressure comes from all directions at once, from the perspective of the individual bodies, and creates a state of equilibrium (of sorts).

The explosion broke up a large mass into much smaller ones. Space, within the galaxy, is really nothing more than the smaller particles spreading out in ever thinner and thinner layers. In other words, matter and space are manifestations of the self-same entity.

The implications of this theory are as immense as space, itself. Matter and energy become two distinctly separate entities with very different properties. Nonetheless, they are like chameleons which can imitate one another.

An explosion, while composed of finite particles, behaves as infinite waves. This gives the galaxy a distinctive sound as it resonates throughout the universe.

A symbiotic relationship exists wherein energy can only travel where there is matter. This insures that each galaxy can develop independent of the other. In a manner of speaking, this keeps our system from bleeding off into space at too rapid a rate.

Calculating the precise hows and whys of our galaxy should be relatively (sic) easy. But what about the Big Bang? Can we explain that in terms which are more rudimentary to our palatial appetite for comprehension?

In the next chapter, I offer my best guess scenario. There are many possible variations of it. Most of those are inconsequential, and would be better served as classroom experiments.

THE BIG BANG BOOMS

Picture, if you will, a huge monolithic structure floating through space. For ease of understanding, let us say that this huge mass is the Earth (without its atmosphere).

Now we cover the whole planet with an immense blanket of methane and nitrogen. This whole ball is quite cold---close to absolute zero.

To further simplify this, visualize this massive conglomeration with several moons of solid ice. These moons are circling the planet at a nice leisurely clip. Let us give this hypothetical planet a new name. Let us call it Jupiter.

Just to make the experiment more interesting, let's add three similar sized, and similarly constructed, planets to the experiment. We shall call them Saturn, Uranus, and Neptune. Yes, of course, we shall put all four into orbitals around a giant orb we'll call Sol.

Aw, the hell with it! Let's just use our solar system as the example. That way, all we have to do is to introduce some fictitious catalyst. Fair enough?

Somewhere, in the vast darkness of space, there is an asteroid which we shall call the XP5000. The XP represents X-tra Potent. And the 5000 denotes that it is 5000 miles wide. It might also mean that it is traveling at a velocity of five thousand miles per second.

The XP5000 is 50,000 miles long and is traversing through space in a straight (sic) line. XP5000 is an icicle of solid oxygen. It has only one mission in life. It is compelled to plunge deep into Jupiter's heart like one of Cupid's arrows.

As XP5000 pierces through Jupiter's 'atmosphere,' it begins to warm up. At the same time, pressure on Jupiter's surface increases because of the tremendous pressure on it. An explosive fire ignites as the tip of this galactic arrow plunges deep into Jupiter. But that is only the beginning.

XP5000 plunges deeper and deeper. Monstrous explosions are fiercely ripping the planet apart. Then the worst-case scenario happens. The icicle strikes the molten hot core of the planet. The resulting explosion is unimaginable.

Jupiter has exploded into a boggling hail of molten, fiery, globules of all sizes. One of these globules strikes, let's say, Neptune, and it bursts into a fiery inferno much like our sun. In the meanwhile, some of the larger chunks strike the Earth. Some fly into the Sun.

Earthlings are powerless to stop the carnage. All that we can do is to watch. And we do so with the fearful realization that the end is at hand. Disaster looms overhead.

Sure enough, some immense chunks of the stricken planet fall into the Sun. Huge solar flares leap higher and higher; more intense than any we have observed in all of man's history. At this point, whether we know it or not, we have approximately eight minutes in which to say our most fervent prayers.

We are aware, from previous experiences with large solar flare-ups, that these eruptions cause thousands of volts of electric current to flow in the very ground that we stand on. Those events had only taken place for a few seconds. And in fairly localized areas. Our newest flare-ups last for several hours, and cause tens of thousands of volts to flow in virtually every region on Earth. Few people, if any, survive the onslaught.

As the massive electric currents surge through the earth, they alter our magnetic infrastructure. More than likely, this disruption causes the Earth to topple over as its poles reverse. This, in turn, causes the Earth's oceans to wash over all of the Earth. And the Earth wobbles like crazy.

Over the course of thousands of years, amphibians will remember to grow legs so that they can crawl out of the murky water. And apes will, once again, swing from trees.

This was, of course, a scaled down version of a Big Bang. Was it entirely nonsensical? Not at all.

A few years ago, scientists peered skyward and announced that an unprecedented event was going to take place. It was not so much that the event had never happened before, it was just that, on those rare prior occasions, the predicted event was never observed by scientists. And the whole world clambered to watch as a very large meteor slammed into Jupiter. The big event, thankfully, fizzled.

What we were not told was that these educated fools were afraid that Jupiter was going to ignite upon impact. Can you even imagine living in a solar system that, at least temporarily, had two suns?

I had the awe-inspiring privilege of living at a time when Haley's comet soared through the solar system. In addition to being awestruck, I felt the fear that most lesser creatures must have experienced at seeing such an impressive occurrence, or abnormality.

Sadly, the comet blasted through all too quickly and we are left with vivid memories of a giant ball of dirt with a trailing tail. And we were saddened by the realization that it will not return for another seventy-five years.

Very recently, we have learned that two far-away galaxies are in the process of colliding. How long will it be until somebody in that system rises to ask: "Did this all start with a Big Bang?"

A LOOK AT E=MC^2

Thanks to a famous man named Albert Einstein, we all have learned to look at the world around us in a very different light. Indeed, a most peculiar and elusive light.

When Einstein introduced his Special Theory of Relativity in 1905, the vast majority of the world simply yawned and then set about ignoring him. Of the remaining few, most were content to lambaste him as a charlatan, a curious little Jew who narrowly escaped confinement in some loony bin, or an outright fraud. To these intellectually deprived denizens of science, Relativity simply could not exist.

Many of Einstein's colleagues refused to delve into the metaphysical realm of Relativity because they did not understand it. And, of the remaining few who did comprehend it, at least on some level, very few dared to wrestle with the complicated equations which the theory represented.

For the better part of ten years, Albert himself would grapple with the complexities inherent in Relativity. Only then could he release a more complete, and updated, version of the theory. He called this version the "General Theory of Relativity."

Through a stringent series of discourses, discussions, and lectures, Einstein's latest version of Relativity was greeted with a tremendous amount of admiration. Even the average citizen (worldwide) got into the act as people squared off on opposing sides of the theory.

As amazing as the theory, itself, was the manner in which people had been duped into embracing it. We all like to think that we are as smart as, if not smarter than, the next guy. When the news was released which professed that less than a dozen people in the world understood the theory, would-be geniuses intimated that they, themselves, fully understood the theory.

Human nature had dictated that men of meager intelligence had to attempt to trick their fellow man by merely suggesting that they understood Relativity. Honest men, many of whom were much smarter than their boastful contemporaries, adhered to the truth. That is, they lacked the training or intuition necessary to perceive such a robust theory.

Man's vanity kept Relativity alive in those early days. Man's vanity continues to keep Physics entombed in a kind of visionary dark ages. In fact,

I have, personally, lost count of how many times I have encountered a stone wall of opposition and bigotry. Do you have any idea how many science institutions steadfastly refuse to consider anything which refutes the current definition of Relativity?

Science has not bogged down because of a lack of intelligence. It has bogged down because of narrow-mindedness, tunnel vision, and an unshakable belief in something which, by all accounts, simply does not work. What?

Whenever you talk to researchers, they will tell you that Relativity works well for some things. But it does not work well for others. Where *I* come from, an equation which does not work in all instances, is an incomplete, or erroneous, equation which clearly leaves the door open for a better theory. Yet, the thugs who stand guard over the various science journals, institutions, and what-have-you, refuse to consider anything not in accordance with a hundred year old theory (Relativity)...even though they will admit it is more philosophical (hypothetical) than factual.

How stupid is stupid? Science has, historically speaking, only made advances when somebody has come along and proven everybody else wrong. In every instance, the collective will of the elitist few has steadily strangled the opposition and refused to open the cognitive door to knowledge. So, whenever I encounter staunch opposition to change, I must question the wisdom of the censors. And I find them greatly lacking.

Remember that old adage? Come on; I know you do. Remember this: Those who ignore history, are doomed to repeat it.

Einstein had his fifteen minutes of fame. It is time to move forward. Question is, where do we move to?

One approach we can take is to dissect Albert's most famous equation. On one side of it, we have an unknown quantity of energy, which we call 'E.' On the other side of the equation, we have another unknown quantity. We call this one 'M' for some measure of mass. The only variable known to us is the speed of light (C). And we square that. Thus, the whole equation is scalar, and amounts to little more than a hypothetical acceleration of 'M.'

That's it. Einstein's most famous equation is for hypothetical acceleration of an arbitrary unit of mass. There is nothing mysterious or profound about that. In every acceleration, be it negative or positive, there is a corresponding change in energy, which energy is equivalent to that which is needed to effect the change in acceleration. More specifically, $E=MC^2$ denotes a change in light energy (quanta).

Interestingly, Einstein's equation(s) work very well in the miniscule sub-atomic world, but not so in the macro-cosmic. The explanation lies in the fact that such high speeds are only possible in the one-dimensional, and two-dimensional, worlds. The three-dimensional world is much larger and speeds are, correspondingly, slower.

It is important to keep in mind that these laws are only applicable in a closed system which contains curved space. Our galaxy satisfies both of those requirements. Once we leave the confines of the Milky Way, the rules change.

In true space, the space which lies between galaxies, the speed limits approximate infinity (for light). What this does to the ratio of mass to energy is anybody's guess. Nonetheless, you can be sure that it is extreme.

Just as dimensions have speed limits, at least within the parameters of curved space/closed systems so, too, do moving bodies. Every body which moves, moves in six directions simultaneously. Motion, therefore, is a net force.

For our purposes, dimensions do not exist. We can deal with them in theory but, in our Universe, everything behaves in three-dimensions. On larger scales, it is not practical to deal with one-dimensional matter...although we do tend to view everything in two-dimensions.

The Unified Field, which Einstein so heartily pursued, was within Albert's grasp. Shunning obscure notions of dimensions, and the idea of a light with constant velocity, we are left with a very simple formula. The size of a given unit of mass determines its maximum speed. This is equally true for things as small as photons, or things as large as our galaxy.

Now we come to the exceptions to the rule. Angular momentum. Velocity is a net force, in a specific direction away from a given point (for our purposes). Got it?

Let's look at a sphere. For simplicity, let us call this sphere 'Earth.' Our sphere has an equatorial circumference of just over 24,000 miles. Earth rotates on its axis at just over one thousand miles per hour. Earth also revolves around the Sun at slightly more than 66,000 miles per hour.

While all of that spinning is going on, our galaxy is spinning, too. Since I lack the data, not to mention the computational skills necessary to manipulate that data, I will leave it to the brainiacs to calculate the net force of all of that spinning. Moreover, I will leave it to them to calculate what affect each body has on the other.

Aside from offering scientific proof that Astrology is founded in reality, what is the point in all of this spinning? Scientists, including the illustrious Mr. Einstein, have been acutely aware of the effects of motion within motion. As a matter of fact, Einstein helped to improvise the gyroscopes which the Germans needed to utilize in order for their U-boats to navigate accurately during World War Two.

If we divide our sphere up into 360 degrees, we see that only 1 degree goes in the same direction as the net force. This means that 359 degrees are in directions contrary to the net force. The consequence is that we are more cognizant of the Earth's rotation upon its axis than we are of the speed with which we orbit the Sun.

Gravity becomes a consequence of both the Earth's rotation and its journey around the Sun. Add to that the motion of the Moon, as well as the motion of the Milky Way, and you can derive at a precise formula for calculating gravity.

So what does all of that have to do with $E=MC^2$? Everything in our galaxy, including atoms, is spinning. Everything also has a net force, or velocity. It seems rather strange to me that a formula which excludes these factors could be deemed accurate. Therein, we can easily see why Einstein's famous equation works well for some things, but not for others. As we shall soon see, in the coming chapters, there are even more problems with $E=MC^2$.

EINSTEIN GOOFED

Albert Einstein was, without question, one of the greatest Thinkers of the twentieth century. He was, also, one of the most misunderstood.

People tend to think of Albert as a pioneer who came up with revolutionary new ideas which had never been thought of. Truth is, most of the various components of Einstein's theories were already in existence. For instance, "quanta" was a concept introduced by the illustrious Max Planck; a full five years before Einstein.

Even the notion of "relativity" already existed. It had been postulated by men such as Poincare, Lorentz, and others. Again, years before Einstein came along.

Essentially, Einstein merely connected the dots and released his findings in abstract mathematical forms. Hence, his infamous $E=MC^2$. But Albert's thinking, like that of his predecessors, was flawed.

Early in his career, Einstein attempted to prove the existence of something scientists called the "ether." This ether was supposed to be some unseen substance in outer space which would allow, or enable, light to propagate from one end of the Universe to the other. Without it, they speculated, light would not be able to move from one place to another.

By the time Albert formulated his theory of Relativity, he had concluded that the existence of the ether was so insignificant and inconsequential as to render it unto the realm of "who cares?" Oh-oh!

In my humble opinion, "proof of the existence of the ether was right under his nose. In fact, it was the ether which was influencing Albert's theorems."

I suspect that the ether runs at, or very near, the speed of light...but, in the opposite direction. "Think of it this way: A light quanta travels to its destination, drops off its cargo, and then returns towards the source to replenish."

For more than a century, scientific researchers speculated that light had no mass. The photon, in particular, cannot have mass because, if it did, it would smash us to smithereens. I assert that a state of equilibrium exists which nullifies the potentially harmful affects of light quanta.

Einstein spent the greatest portion of his life searching for the Holy Grail of Physics: A Unified Field Theory. According to my calculations, "It is ironic that Einstein abandoned the one thing that would have enabled him to piece the puzzle together. The ether is real!"

Because the ether is a component of a system in equilibrium, it is virtually undetectable. If we push X amount of Ether ahead of a moving object, X amount of Ether will move behind it in the same direction of travel, and at the same magnitude.

"The Ether exists and I can prove it." I can say without hesitation. "And I will explain how you can prove it, too."

The method of proving it is the topic of another chapter. So I will leave it at that for now.

THE ARGUMENT FOR ATTRACTION

It is pretty easy to argue the case for attraction. We look at phenomena such as the forces exhibited by magnets, and we feel that we have no choice but to conclude that the force of attraction exists. And, yet, I am still not convinced. Not even the apparent effect of the Moon upon the tides can sway me.

However, I am perfectly willing to play the Devil's Advocate and argue in favor of attraction. Let's see how that pans out.

The forces of attraction and repulsion imply a relationship such as negative-positive, push-pull, or any similar force relationships. Attraction, Negative, and Pull, all suggest a void wherein atoms can migrate. That is the simplest scenario.

In this 'void' scenario, we have microscopically big canyons in which certain atoms can fall. Still, in order for such a scenario to exist, both the entrance to the 'canyon,' and the atoms spilling into it, must be smaller than other atoms. It is plausible. And it lends credence to the possibility that negative numbers could have physical manifestations within our Universe.

Obviously, if we cannot advance a satisfactory argument favoring the force of attraction, even with the aforesaid simple scenario, then something else must be going on.

Numerous scientists have attempted to explain the force of attraction, at least in some gravitational sense, throughout the ages. To the ancients, magnetic attraction, via lodestones, was magic. To later investigators, such attraction was the result of properly aligning the bipoles within a given magnetic unit. Modern researchers toy with the idea of atoms throwing dust at one another.

My personal favorite is String Theory. Envision, if you will, my 'advanced' version of String Theory. Let's call it "Z Best Theory."

In Z Best Theory, we take microscopic pieces of string and hand each end to a person standing on an atom. Thus, two people float around playing a fastidious game of tug-of-war. All day. Everyday.

Just as their real-life human counterparts, these little guys get tired and they drop the string. A new (rested) opponent then picks up the trailing string and a new battle ensues.

This silly discourse is, logically, repulsive (sic). Nonetheless, it does open the doors for serious speculation. If we eliminate the tiny string holders, and substitute some, as yet, unknown holders/fasteners, the idea of tethered atoms is not nearly so preposterous.

Carrying the concept of tethered atoms a step further, we can assign differing qualities to each of the tethers. One can be stiff like wire. Another could be flaccid like rope. Still another could be flexible like a bungee cord. Perhaps we can call these thrill seeking atoms "bungee jumpers!"

To this whole conglomerate of tethered atoms, we can add in other factors. Again, these qualities would only apply to the tethers. For instance, we can vary the lengths so that two larger atoms could only 'duke' it out if they encountered a longer tether.

On the surface, it would seem that this fanciful version of string theory would be readily acceptable to even the most staunch supporter of either string theory, or of the force of attraction. The inherent problem is, how do we define the contact points for the strings? Are they held in place via ancient magic? Or is there a force of attraction at work in ways that are unidentified by us?

I must say, with some relief, that I have failed to advance a proper argument favoring attraction. It is possible that I failed because I was wanting to fail. On the other hand, maybe I failed because attraction is a myth. Whatever the reason, I'm sure that you won't want to miss the next chapter.

EINSTEIN'S SECOND MISTAKE

We are all so conditioned to believe that everything in the known Universe is symmetrical that we obscure our vision and our intellectual capabilities. This symmetry is a logical conclusion based on what we can observe. We see north-south, up-down, negative-positive, etc., etc. But symmetry is an illusion.

Each pairing is merely the opposite ends of the same thing. Instead of being two disparate functions, these are all the same. But they appear opposite because we view them from opposite ends of the event(s).

In the most revolutionary discovery of the twenty-first century, I have concluded that, "Push" is the only force at work in our Universe. "Pull" is only an illusion which simply does not exist in the three-dimensional world. It is an elementary concept which eluded Albert Einstein in that great man's quest for a Unified Field Theory.

Back in Einstein's day, scientists were too busy squabbling over the existence of the atom to take a closer look at forces. Only a handful of men were brave enough to actively promote the atom. I predict that my theory will meet with the same initial resistance. Still, I am confident that my conclusion will be vindicated.

To illustrate the degree of opposition I expect to encounter, I point to another belief that has long been held onto by Physicists. That belief is that nobody "outside of the profession" will ever receive a Nobel prize for his work. Wanna bet?

If I am correct, then I have given researchers the missing ingredient for the Unified Field Theory. If that theory is the chief goal of Theorists, how will they ever justify excluding an "outsider?"

Right, or wrong, one thing is for certain...life just got a whole lot more interesting.

REDEFINING GRAVITY

Early on, I had come to believe that I was on the verge of proving that Einstein goofed as he assembled his theory of Relativity. I had cogitated the idea of eliminating the force of attraction several years ago. And I knew that it contradicted much of Einstein's work. Just how true that was did not surface until I made an even more startling discovery.

Having formulated my "Push" theory, I went in search of proof which would substantiate, or refute, the idea. I looked at the moon and realized that it could not be attracted to the Earth. This, however, was not an entirely knew idea to me.

I had always been fascinated by magnets. I knew that "like" poles, or charges, repelled. Opposites attracted. It was these simple rules which I always applied to the moon.

The moon reflects sunlight. That could only happen if the moon absorbed enough energy from the sunlight to become saturated. The Earth, also, reflects sunlight. Again, this could only occur if the Earth had become saturated, too. Obviously, if we have two bodies which contain like charges, then they are repulsive to one another. Extremely logical.

So far, that only utilizes classical Physics. Anyone with a minimal of intelligence can see that.

Like millions of other people, I embraced Einstein's Relativity and applied it to the Earth-Moon relationship. I had no problem with putting both of them inside a curved space that wrapped around them like taut cellophane.

From there, I toyed with the idea that space was compacted in the areas all around each orb. It is a hypothesis which I only slightly veer away from today.

In my mind, there are two types of space (perhaps more). There is the space that needs to exist just as an artist's canvas. Without this stationary space, there is no 'art!'

The other space, the one I call the Ether, is much like a succession of gears. It moves with matter. This space would best be described as semi-rigid. And it moves at right angles to primary space. It is this space which curves and gets compacted (denser) in places of greatest activity.

I would almost be tempted to call these two spaces electro and magnetic. In that way, I could assign them very real attributes. Moreover, such definitions would alienate them from the dust particle space which permeates all regions of a closed system.

It was very easy for me to see that the moon was not attracted to the Earth. And vice-versa. Still, I wondered about the existence of any other proofs.

Introduce Pluto and Charon. One is (or was until recently) called a planet; the other, a moon. The fact is that many Astronomers do not, and have never, considered Pluto to be a planet. I add myself to that list…kind of.

Pluto and Charon are an interesting phenomenon. They are, essentially, the same size---more or less---and go through space tumbling around one another. It is a beautifully strange dance at the outer fringes of our solar system.

I had found my second clue. Pluto and Charon could join together…if they were attracted to each other. But such a marriage would never see a honeymoon. They can never join as one.

Without a doubt, if a force of attraction existed between the two, the curvature of space around them would add to the attraction. I had found my second example in support of push.

One thing that I had, inadvertently, overlooked in my efforts to resolve Einstein's mistakes, was the simple fact that I was really trying to define gravity. And that was the precise instant that the gravity (sic) of the situation hit me. I wasn't just trying to define gravity; I was rewriting the fundamental equation with which so many learned men had devoted their entire lives.

Isaac Newton had defined gravity as a force of attraction. Every great man who followed, blindly accepted Newton's definition as gospel. Whereas I had merely intended to expose the error in Einstein's attempts to link gravity to relativity, I had really, in actuality, gone all the way back to Newton.

And Newton was wrong. Boy-oh-boy. His formula for calculating the force of gravity had worked so well (for many purposes) that nobody ever thought there could be another answer. Until now.

Specifically, Newton said that the force of attraction between two bodies (usually planets or moons) is directly related to the mass of those bodies and inversely proportional to the distance between them. What that meant was that bigger bodies would have more attraction than smaller bodies. Just what that force was, nobody could say for sure.

One of the most obvious problems inherent with Newton's hypothesis was that it did not make any difference what these bodies were composed of. All that mattered was their physical size.

Think about that one for a moment. A giant ball of steel would attract an equally massive ball of glass. In fact, according to Isaac, the two objects could be any shape and, therefore, did not need to be round, oval, conical, or spherical. They could be any shape!

Well, you really don't need for me to tell you how crazy such a postulate is. So I won't. However, I do feel a certain compulsion to explain the most plausible reason why Newton's theory was so widely accepted---particularly by advocates of the Relativity theory.

Einstein speculated that gravity was a consequence of space warping. In other words, the larger the body, the greater the space around it warped. That was a very noble attempt at explaining the phenomena. And there may be a thin element of truth to it.

Now you must be wondering just how I propose to explain gravity in such a way as to eliminate the force of attraction, and still allow for the components of Newton's theory. Actually, that is quite elementary.

Our galaxy is the result of a massive explosion. No, not the Big Bang. Much, much, smaller. But large, nonetheless. And the fact that the Milky Way is spiraling we can deduce that the explosion was the result of a major collision between a couple of suns, or so. This means the suns collided slightly off-center to one another.

Be that as it may, the force of the event is still contained within our system. My best guess is that explosions in space occur in slow motion...and last for a rather long interval of time. That is, relative to us, it is a long event.

The force of the explosion, possibly as some form of heat wave (cold wave?), bounces off of the outer walls of the galaxy...and everything within the galaxy. Because it is a closed system, there is an equal magnitude of force being distributed throughout the galaxy.

Though this force is spread equally, it does not act that way. Let me give you an example. Let's say I put a golf ball in the very center of a pressure cooker and seal it in. As I pump air into the cooker, the golf ball should remain stationary.

Now, let us open the pressure cooker and place a second golf ball inside. We seal the cooker and begin to add pressure. Eventually, the two balls will collide.

Next, we put a baseball in the center of the cooker, place a golf ball somewhere in the vicinity of the baseball, then add pressure. Only this time, we sit the apparatus on a Lazy Susan and spin it around. If we do it right, we should observe the golf ball 'orbiting' the much larger baseball.

Given enough time, it would not matter where we placed the two balls originally. Over time, the much larger ball would always find its center of gravity.

Two bodies in space are under the same sort of force. If we draw a line, a straight line which begins at the outermost edge of the galaxy, passes through both bodies, and ends up at the outermost edge of the other side of the galaxy, we will make a simple discovery. The space between the two bodies is shorter than the space between the bodies and their respective galactic walls.

LOOKING AT DIMENSIONS

One of the greatest blunders of the twentieth-century was placing *time* in the fourth dimension. We have all been taught to place things in their order of importance. This might explain why the fourth dimension has been so badly downtrodden or misunderstood.

According to Atkins, "Time is everywhere. If time did not exist, nothing would exist." Touché!

Because time permeates everything in the known Universe, it should be relegated to its rightful place in the first dimension. Next would come space because it is the stage from which all actors act. Third would come light. Fourth would be electricity. Fifth would be magnetism. And sixth would be matter. Why?

Time is infinite. Space can be infinite or, as many suggest, "finite, but without borders." From the infinite, we work our way on down the speed limits.

The first observable phenomenon, which has an observable speed, is light. Light is a one-dimensional entity which sets the speed limit of the first dimension at about 186,340 miles per second (mps) in our galaxy. Atkins intends to prove, in another paper, that the speed limit for light is near infinite outside of our galaxy.

Going on, electricity is at the border between one-dimensional and two-dimensional. Electricity travels at about 186,000mps. At the other end, lies magnetism. Magnetism is at the cusp between the second dimension and the third. It travels at a much slower speed than electricity. Both electricity and magnetism are two dimensional entities.

Matter, or three-dimensional objects, are much slower. These speeds are regulated by states such as Gas, liquid, or solid. It is important to remember that, in no known dimension is there anything that is purely at rest (relative to the known Universe).

From the aforesaid facts, according to Atkins, we get a dramatically improved picture of the relationship of energy to matter. Effectually, energy is matter in motion, and matter is energy at rest (but not zero).

BLACK HOLE OBSERVATIONS

In mathematics, we often utilize nonsensical expressions which are then applied to explain hypothetical situations, or events. Negative numbers, and fractions, immediately spring to mind. And the saddest part is that all of us have developed thought processes which enable us to grasp these abstract concepts.

Let's say that I have a stack of ten silver dollars. To this, I wish to either add a negative five silver dollars, or I wish to multiply by the fraction 1/2. What do I wind up with?

We, of course, conclude that the outcome is that I now possess half as many silver dollars as I started with. The reality is, there is no such thing as a negative five silver dollars. Likewise, if I multiply 10 silver dollars by 1/2, I do not end up with five silver dollars. What I do end up with is 10 silver dollars that are now 1/2 the size they were.

Let's look at another example. Let's say that I have a silver dollar. To this, I wish to add a negative twenty silver dollars. In colloquial math, we deduce that we have a net result of -19 silver dollars. It is an abstract result which cogitates such distressful notions as debt, deficit, and being in the red.

Okay. We've got the basic idea. So what? Well, now we substitute. Instead of silver dollars, let's say that I have a Moon. To this, I add a negative 20 moons. Now what?

If, indeed, there were a force equivalent to a negative 20 moons, we might expect to see a black hole where our Moon used to be. Strangely enough, we should expect to get our Moon back by adding 20 moons to the black hole. Possible?

Who knows? Such hypothetical preponderances may occur in reality. Or they may just be fanciful musings. If they exist, then it is possible that parallel Universes exist. But not in the conventional manner of thinking.

So much mystery and hoopla has been advanced, as regards the black hole, that it almost seems criminal to demystify them. Irregardless, science must move on.

Every galaxy in the Universe is spinning. This signifies that the space within those spinning clusters is curved. Light, traveling throughout these galaxies, is moving at a rate of speed relative to the rate of the rotating galaxy or, if you wish, at a rate proportional to space density/curvature.

I will come back to this spinning, as well as its effect on light, in the next chapter. Let's suffice it to say that all of this spinning slows down light.

The point to be made here, is that there exists two distinct types of space. We have a closed system in which space is spinning (our galaxy). And we have an open-ended space which is at rest (relative to our galaxy). All scientists know that things behave differently when they are at rest than when they are in motion.

In the case of the spiraling galaxy, light travels slowly through the curved space. When light exits curved space (our galaxy), and enters the open-ended space beyond the Milky Way, it rapidly accelerates. A black hole, as well as all black matter, is nothing more than places where curved space has given way to true space.

THE ABSORPTION THEORY

The entire Universe works on a principle called Cause and Effect. It has, at its roots, another principle which states that Necessity is the mother of invention. As a matter of recourse, we could say that necessity lies at the heart of any precluded Unified Field Theory. And the reason for this is simple enough. Everything exists for a reason.

For those of you who ascribe to Biblical intonations, we can turn to Ecclesiastes for confirmation. Everything has a time and a season. And the inference we can draw from this awareness is this: A thing exists for that time and for that reason.

Another inference we gain from Ecclesiastes is that all things are important. Take, for instance, a strawberry plant. For eleven months of the year---nothing. Then in the twelfth month, gloriously, the plant yields succulent red berries which we can preserve for consumption in any of the months to follow.

Point is, for all of our collective intelligence, we can see many things, yet we do not see their purpose. Let him with understanding, understand; him with wisdom, to be wise; and those without a clue...just shut the hell up!

That wasn't very nice. But oh so true. I am so tired of so-called professionals telling us to believe in this or that, and then they admit that they don't know.

Yes, I am taunting the face of hypocrisy. I ramble on, reaching, it would seem, for straws. And I confess that I am ignorant as to all things. In that context, I am as guilty of misdirection as my fellow man.

One difference between what I seek to accomplish, and what the others are doing, is that I am trying to reprogram your computer (brain). I challenge you to buck the trend. All great accomplishments have come from those who swam against the tide of conventional thinking. Dare to be different. Question everything.

One of the things which I have gained a reputation for doing is to lead you down a path which seems to have little, if anything, to do with the topic at hand. The short answer to that is that the complex nature of my mind sends me down all of these paths simultaneous to arriving at a conclusion. The odd thing is, these excursions into unrelated areas often turn out to be highly applicable after all.

The topic before us is 'Absorption Theory.' What can the mistaken beliefs of researchers, and your adherence to those mistaken beliefs, have to do with absorption? For the answer, let's define the Absorption Theory.

In the exciting world of electronics, there is a wonderful process called saturation. Essentially, saturation means that a unit of material has "filled up" with energy and can no longer absorb additional energy. In electronics, this energy is usually a flow of electrons.

Technicians have a pretty good idea how many electrons a given unit can "absorb" before reaching the saturation point. This knowledge enables the technician to construct various circuits which function a certain way when the given unit is in a particular state, or the other. That is to say, as long as the given unit is 'thirsty,' one circuit will remain in operation while the other(s) are not. As soon as the unit is full, a second and/or third circuit takes over and the electrons are diverted away from the unit.

Astrophysicists tell us that every sun in the observable universe exhibits absorption lines which they call 'Fraunhofer Lines.' Basically, these are dark lines which show up in spectral analysis of sunlight. They appear as dark lines because there is no light in them. Think of them as shadows.

According to Physicists and Astronomers, these dark lines are caused by cool gas (high above the sun's surface). They claim that this cool gas absorbs only certain wavelengths of light and, thus, enabling researchers to identify the composition of these gas clouds by the absent wavelengths.

The prevailing theory about why this is so, is that certain elements absorb specific wavelengths. The other side of that equation is that certain elements emit specific wavelengths that other elements do not.

It seems to me that if the whole spectral analysis thing is so accurate, why don't we eliminate the Mendeleev Table and replace it with a color coordinated chart? We'll divide the chart in half. On one side, we place light emitters. On the other side, light absorbers. And, just to top it all off, we'll demonstrate how the light absorbers become light emitters---once they reach saturation point.

Such an absorption theory (table) is not outside the realm of possibility. And it does make a fair modicum of sense. Hmmmmmm...

PROVING THE EXISTENCE OF ETHER

A hundred years ago, it was hypothesized that, if an Ether existed, we should be able to measure it in some way or fashion. Indeed, the notion that a stationary Ether existed fit very well into various theories and equations. Unfortunately, all attempts to prove that an Ether occupies space have failed--- until now!

Convinced that an Ether exists, and equally convinced that light should travel much faster than a paltry 186,340 miles per second, John Hildreth Atkins has formulated a plan that he believes will prove that the Ether is really there. And, almost as a side-bar, John's plan will provide us with a clue as to some, if not all, of its characteristics.

So what, exactly, is Mr. Atkins' proposal? Placing a laser on the Space Lab, and firing it at mirrors on the Moon (or beyond). By pulsing the laser at diminishing intervals (I.E.- one second spacing between shots, down to 1/100 second pauses), we should see an erstwhile response.

Explanation: If light is faster than 186,340mps, and this slower speed is a consequence induced by the Ether, then there is a good chance that the intervals between shots will disappear once the "faster" light catches up to the pulse in front of it. This will occur when the shots occur at a rate faster than the Ether's ability to "mend."

FUTURE TIME

Just a quick note...I was going to write an article/chapter about future time. This being the speed of time and its affect on the way we interpret the universe. It stems from my belief that we are in the midst of "fallout" from a huge explosion and that caused time to slow (as compared to our ability to comprehend). At any rate, I will deal with it at a future time. Smiley face.

ESCAPE VELOCITY

Scientists have done a wonderful job of calculating escape velocities for the Earth and Moon. But why haven't they done the same thing for all things (including the atom)?

My first impulse is to answer my question with a vague generality. I could simply say that such calculations are much too complex. However, in this day and age of computer literacy, it is merely a matter of programming.

If we possess the empirical data, and provided we have the correct formula, we should be able to predict for all objects. But, alas, we cannot. And the reason is simplicity, itself.

Let's take a sphere and half fill it with sand. Next, we fasten a tether to it so that we can sling it around in a circle. We now note that as we increase the speed of our orbiting orb, as it circles us, the sand begins to gather at the outer boundary away from the fulcrum (us).

Do you see the picture? Our orb has an escape velocity. If we spin it around us fast enough, it will fly off into space. But, is speed the only factor? Of course not!

The strength of the tether, the strength of the fastener, and our grip on the other end of the tether, all come into play. Earlier, I referenced that motion was the net force of the total forces at work. I also stated that a moving object moves in six directions simultaneously. I could carry that postulate further and simply state that all net forces are the result of, a minimum of, six forces. Escape Velocity, thus, depends on the mass of the globe (including the sand), the speed at which we accelerate it, the strength of the tether, the strength of the fastener, the strength of our hand, and the forces which make up the ambient.

After our orb has reached the saturation point (I.E.- the point where it cannot take on any additional energy), any additional energy (you can substitute mass) will cause it to 'escape.' With this escape there is an accompanying moment of hesitation, which is followed by a resulting inertial acceleration. In other words, once the threshold of Escape Velocity has been compromised, one of four things can occur.

According to Newton's laws, our orb will continue to accelerate until it is acted upon by an outside force. Therefore, our four choices become: It can stop, slow down, maintain a constant speed, or it can speed up. Again, these actions are contingent upon ambient, or lack of ambient, forces.

THE BIG PICTURE

As much as he was able to at the time, Albert Einstein attempted to build a mental imagery of the Universe around him. Such a feat was tantamount to looking at a microbe with a microscope---not impossible, but darn near.

What dear Al ended up with was a model wherein the Universe is unlimited, but is finite in volume and in all directions. We can liken this concept to a balloon which provides us with a visual two dimensional surface within a three-dimensional system. If we were on the surface of that balloon, we could walk forever, in any direction, and never encounter a boundary. The balloon, like the Universe inside of Albert's mind, contains a set volume within a clearly defined area.

When Einstein pieced together this particular concept of the Universe, he was under the impression that the whole Universe was curved and, therefore, the light from a distant star, light that had sufficient energy with which to make such a lengthy journey, would be visible in two positions within the parameters of his Universe. The light would be visible in the relative location of the origin of the light and, also, in the opposite position.

This dual aspect theory could best be illustrated by the following analogy. Let us draw a point on the surface of our balloon and label it a star. Now, we can draw a line around the circumference of the balloon. This we can label the path of a single ray of the starlight. Next, we can point to any position along that line and we observe that, no matter where we select our imaginary point of observation, there is a line in front of us and a line in back of us. This, allegedly, confirms that our starlight is visible in two opposite directions.

The only drawback to the above-referenced experiment is that such hypothetical thinking does not adequately represent infinite space. However, the balloon is an excellent purveyor of a galactian model. Think of it in terms of the balloon being representative of our galaxy, the Milky Way, and start to spin it. Have a couple of million friends or associates come over and hand each of them a balloon. Tell them to start spinning their balloons. Now, we have a very, very, good model of the real Universe around us.

Fact is, we live on a planet that is spinning inside of a galaxy which is also spinning. We, therefore, must deal only with moving forces. Everything inside of our balloon is moving in an arc. Everything outside of our balloon moves in a straight line.

If you have ever looked at the night sky and wondered how the light from a star, in a galaxy which is a million light-years away from us, actually reaches us, you now have your answer. Light, which travels in a straight line between galaxies, travels at, or near, infinity. But how?

In our galaxy, space is curved. We can use Newtonian physics to calculate the effects of inertia, angular momentum, etc. Those with extreme intellect, can see that curved space is compacted (dense). Light travels through dense mediums at a much slower velocity. Moreover, in open (non-curved) space, no dimensions exist.

Let me put that into a different perspective. Quantum physics tells us that a three-dimensional body, with a given mass "M," traveling at one hundred miles an hour, will do more damage than a two-dimensional body with the same mass and speed. This is because when the two-dimensional body (essentially a photograph) encounters the three-dimensional body (all things equivalent), the two-dimensional body's mass will be redistributed in the three-dimensional. The result is a lessening of 2-d's force.

Let's look at a classical equivalent of that example. Let's say that we have three batteries. One is a four volt, one is an eight volt, and one is a twelve volt.

Now, let's suppose that we have three dimensions. Each dimension is capable of holding a maximum of four volts. Let's start out with the four volt battery and place it in the first dimension. It zips along until it encounters either the eight-volt, or the twelve-volt, battery sitting comfortably in its own little sector of the universe.

At the time that the two batteries collide, several things can occur. The two batteries can zip off in new trajectories, they can fuse together, they could altar one another, or they can unite as one (3-dimensional) battery.

One thing to bear in mind about the last postulate is that the 2 combined batteries are not equal to the singular twelve-volt battery. The united batteries can either be greater than, or less than, but not equal to the singular battery. What?

To better visualize this, let's take an excursion into carpentry. In carpentry, it is often necessary to "laminate" boards to one another. In some instances, this creates a compound that is, by itself, weaker than a single board of similar mass. In most cases, the process of lamination is to reinforce, or to make stronger.

The point of this whole exercise is to show that a three-dimensional object has more energy than a two-dimensional, the two-dimensional has more than a one-dimensional, and the one-dimensional has more than space (0-dimensional). For the record, curved space is two-dimensional, light is one-dimensional, and plain old space is 0-dimensional (for our purposes).

Therefore, our one-dimensional light slows down in 2-d space. It can only travel at 186,282mps...until it enters non-curved space. Since 1-d is a force greater than 0-d, light can now travel unimpeded. And continues accelerating at C squared (exponentially).

Our next problem appears when we attempt to travel. Three-dimensional matter cannot travel as fast as 2-d, 1-d, or 0-d, matter. But, even before we can address those problems, we have other hurdles to clear.

Virtually every human being alive today is aware of a little phenomenon called Escape Velocity. There is an escape velocity for the Earth. There is an escape velocity for the Galaxy. Matter of fact, there is an escape velocity for every single atom in the Universe. But let's not hasten there right this minute.

The question now arises, "Why do we have escape velocities?"

Aside from the fact that God designed it that way, we have stumbled upon a fourth dimension which we shall call "Time."

The problem that we face when dealing with the concept of time is that we do not have a uniform definition of the concept. Indeed, for most of us, "freezing" time implies halting forward progress which, in turn, implies that time is motion. Oh-oh!

Einstein's famous equation banks on two postulates. First, the notion that the speed of light is constant. And second, time itself, as related to light, is constant. While the speed of light, within the framework of our "closed system" galaxy, is close to constant, time, most definitely, is not. Let's look at some practical examples.

Let's say that we get into a car and we travel at the rate of one mile per hour (mph) down a given highway. It is a hot day and we have the wife and kids in the car with us. I can pretty well guarantee you that, at the end of one hour, your family is going to make your life more miserable than it already is.

Now, let's return to that same stretch of road. Instead of one mile per hour, let's go 60mph. At the end of one minute, we have already traveled the same distance as our earlier venture. Better yet, is the fact that nobody in the car has gotten cranky or miserable.

The argument arises that we have merely accelerated time. That is horribly erroneous and grievously flawed. For time to have been accelerated, we would need to replicate every event within the 1mph drive. Did we overheat? No. Did we get bored? No. Do you see the point?

Time is not strictly about motion. In fact, if we froze the Universe, time would continue because the universe would still be there. Why? Because potential energy has a time coordinate (coefficient).

Some theorists will argue that if the universe were to freeze, it would cease to function (exist) and time, would also, be nonexistent. Nice theory. But rejected and rebuked by both the Conservation of Matter, and the Conservation of Energy.

An interesting hypothesis arises out of this abstract thought. If we did have a situation wherein the Universe froze, it becomes obvious that we would need the addition of a single moving particle in order to put it into motion again.

I use this abstract to make numerous observations. First, the particle would be the reverse equivalent of a Universe Escape Velocity. The calculation for such a particle would be of a magnitude comparable to the alleged Big Bang theory.

Ignoring the speculation about where such a particle may have come from, as well as what may have propelled it, let us look at a more likely scenario. Let's consider Brownian Movement.

As we are watching the jitters of suspended molecules, which are reminiscent of people who have imbibed too much caffeine, we realize the huge energy potential which is just waiting for liberation (a direction). The liberating particle, say a beam of light, enters and these Brownian Ballerinas spin away.

Physicists know that a good explosion depends on the energy potential, as well as the time (duration) of the release of energy. The nearer the magnitude the potential energy can come to the saturation point of time, the greater the magnitude of the explosion.

We can go back to our car experiment and substitute a bulldozer for the car. We drive the bulldozer at the rate of one mile per hour, twenty-four hours per day, for one full week. We push pile after pile of dirt into a humongous mountain. At the end of the week, we have one hell of a hill in front of us.

Now, let's change the time coordinate. We create the same mountain, but we construct it in one ten-thousandth of a second. What do you suppose happens?

Well, assuming it were possible to even do that, we would probably have a crater in the Earth about the size of the Grand Canyon. The reason for that is simple enough. Heat from things like friction, exhaust, and pressure, would not be able to dissipate. The heat would accumulate very rapidly and the net result would be a huge explosion.

GASOLINE ENGINES ARE OBSOLETE

An Oregon man claims that gasoline engines are obsolete and have outlived their usefulness. Moreover, according to John Hildreth Atkins, gas powered motors have been headed the way of the dinosaurs for quite some time.

"Forty years ago," Atkins says, "I built a motor out of magnets and some old bicycle parts. It was cool but, even then, I had a sense that it was obsolete, too."

"There are so many energy sources out there, just waiting to be utilized, that it seems absurd to be fighting with OPEC."

"The United States used to be known as the most technological country on the planet. Now, it seems, scientists don't even know how to tie their own shoes."

According to Atkins, physicists have bogged down in a quagmire because they took the path of least resistance. Instead of searching for the truth, they found it easier to build on the work of their predecessors. For instance, it has long been accepted that all atoms carry atoms. That is a fatally erroneous assumption.

Be that as it may, Atkins frequently wonders why these magnet motors are not in every home. They do not pollute. They do not require the consumption of fossil, or any other, fuel. They are cheap to build, require little maintenance, and just plain make sense.

"Corporate greed is the only viable explanation. Magnet motors provide an alternative energy source which would devastate oil companies, as well as commercial power suppliers."

Therein lies the rub. The almighty dollar has replaced good sense. And we are all paying with our health...not to mention our pocketbooks.

GLOBAL WARMING

If the solar system is an accurate depiction of some force of thermodynamics, then we have another factor to consider when examining possible outcomes of global warming. I am, of course, referencing the reversal of the Earth's magnetic poles.

How can I be so sure that global warming is, in reality, a cause of such a catastrophic event? Other than my keen intuition, science will bear it out in several ways. Some of these are reliant upon classical physics. Some are contingent on (new) theoretical physics.

To begin with, we know that heating up an iron magnet will stop its magnetic properties. Heat, as we all know, affects air flow, air flow creates (static) electricity, and electricity creates magnetism. Each is a logical consequence which we easily accept as "fact."

Now we come to the theoretical side of this equation. It is this equation which is suggested by the solar system, itself. More specifically, the motion of the planets.

Looking at the first four planets (Mercury, Venus, Earth, and Mars), we observe that these four planets rotate on their axis at a much slower rate then their four larger brothers (outer four planets). Unfortunately, there are many reasons for these discrepancies.

First, the inner four are moving around the sun at a rate of speed which is much faster than the outer four. Friction might, therefore, be the deciding factor.

Second, the inner four are many times smaller than the outer four. We know that smaller objects (bodies) can, and often do, move faster than their larger constituents.

Third, the surface temperature of the inner four is considerably higher than that of the outer planets. By my estimation, it is this temperature differential which is causing the planets to rotate on their axis.

So? Well, I could be wrong, but it seems to me that thermodynamics plays a pivotal role in a planets movements. If so, then, as our sphere warms up, it should be slowing down.

Picture a ball bearing. If it is dry, it does not function efficiently. Eventually, the ball-bearing will get rusty and stop working. However, we can grease up the ball-bearing and it will function very well again.

Planets, and moons, do not seem to function (rotate on their axis) if they do not have atmospheres (GREASE). Cold surfaces, atmospheres, and size, all seem to contribute to the rate at which a planet moves.

One thing which I am looking at is polar ice. My bet guess is that polar ice is the grease which enables the planets to rotate on their axis more efficiently. If so, then as one of Earth's poles loses its polar ice, may cause the Earth to wobble and tilt. This would explain the wobble and the reversal of the Earth's poles.

Scientists say that our moon is moving away from us. This particular movement would be an expected consequence of Earth's slowing down.

It would be easy for me to take the obvious approach and state that when the Earth slows down enough, the moon will disembark on a journey that is far removed from Earthly experience, and leave it at that. But that isn't my nature. So let's shake it up.

I predict that, as the Earth slows down, the moon is going to get further away from us. Just how far it gets is going to depend on several factors. One of these will be global warming. And that just happens to be the only factor which we have any control over!

The next thing that will occur is that Mercury will get pushed into the sun. Venus will begin to rotate faster when the moon gets within orbital distance of it. To balance out the mass equation,

Earth will capture Mars. Deimos and Phobos will go to Venus.

The resulting destruction will be catastrophic.

IN CONCLUSION

I had actually written the afore-stated articles individual of one another. I had no intention, and certainly no idea, that I was about to unite them all into a common thread. That thread, it would seem, is mainly centered around the concept of speed.

It is common practice to say that there are three states of matter. Gas, liquid, and solid. Nothing difficult about that acknowledgment. But let's re-examine it; shall we?

Each of those three states is in reference to three-dimensional matter. With a little bit of a nudge, we can easily see where gas can be classified as an attempt to transmute from three dimensions to two dimensions. In other words, the closer we come to eliminating one of the dimensions, the greater we increase our speed.

As a side-note, at the time that I write this, I am indecisive as to which end of the dimensional spectrum radioactive material would be placed.

My own inclination is to put it at the bottom (slow) end. Perhaps it is so tightly packed that it is collapsing in on itself. But we won't worry about radioactive material at this juncture.

I realize that the natural progression is from slow to fast, from solid to gas, and there is sound reasoning behind this transformation. In its simplest form, we are adding energy. As we add energy, by definition, we are increasing activity and all of that energy (pardon the pun) has to go somewhere.

My earlier declaration was that magnetism is slow (for two-dimensional existence. If gas is also at the cusp, then it stands to reason that there might be some symbiotic relationship between magnetism and gas. Gas tends to be hot and magnetism tends to be cold. Ignoring that possibility, we are still left with an approximate speed limit for that border.

I am sure that many of you have already postulated that if Matter has three states, then, perhaps, energy must, too. And, if you are particularly clever, you will have surmised that each dimension has three (or more) states. And, if you are really wise, you will come to the conclusion that residents of the first, and second, dimensions must be energy...or, at least, more energy than matter.

As speed, velocity, or acceleration increase with the decrease in the number of dimensions, another factor arises. Matter has, suddenly, become energy.

And the faster that a particle moves, the less energy it takes to move it---until you reach its saturation point. Let's hold that thought for a minute.

We all know that it takes a computed infinite energy to move matter at the speed of light. The reality is, you cannot move matter at the speed of light. Matter will convert to energy long before then. It will also shed one, or more, dimensions.

Energy, or the coefficient of energy, is just the opposite. Why? Because energy, in its purest form, travels at infinite, or near infinite, speed. You cannot pile an infinite on top of an infinite.

Now this creates an unusual situation in that it takes energy in order to slow down energy. This energy exchanges for matter rather nicely. That is to say, a very small piece of matter can stop a very large piece of energy. So? What am I getting at?

The point that I am so miserably trying to put across is simply that as light slows down, it increases in mass/force. We can actually derive at that conclusion from another direction. Take the case of our sun, for example. It houses a tremendous amount of energy which it uses to expel photons. Since energy is conserved, whatever force expended by the sun is going to be the same force required to slow the photon.

From the atom, we now return to the cosmos. I bring you back to this point because I have a little theory which I am sure will interest others. I theorize that there is an invisible barrier somewhere between Mars and Jupiter. This barrier only allows bodies under 13,000km to pass through unscathed. Larger bodies are ripped apart.

This invisible barrier is, most likely, the outermost shell of an electron. The barrier is more of a line of protection than a means of containing whatever happens to be inside of its protected region.

I predict that the barrier is closer to Jupiter. In fact, I believe that it is so close that it may even affect some of Jupiter's moons (particularly the larger ones). I leave that to scientists to discover.

I used to tell people that I could provide them with the answer to their question---if they could explain their question to me. This was my way of bypassing the education that I knew I was lacking. For instance, I had never heard of Brownian Movement and, therefore, had little hope of being able to explain it.

I have since discovered that Brownian Movement is a jittery dance that some solids seem to exhibit while suspended in a liquid. It has become clear to me and, from there, I have derived at a unilateral relationship between energy and matter, and not-so-steady state.

My point is not, despite contrary opinion, to brag about some imaginary intellectual prowess which I may possess. My point is to impress upon my readers a rule which Physicists have long since abandoned. That is: Make sure that you understand what it is you think you are seeing.

Sir Isaac Newton is said to have observed an apple falling to the ground. From that event, he speculated that the moon attracted the earth. And we all know the result of that observance.

Newton went on to postulate that gravity was a force inherent in all things. He defined this force as being an attraction between two bodies. The bigger the bodies, and the closer they got to each other, the greater the force of attraction. Just why the moon did not plummet to the Earth is anybody's guess.

Newton's observations satisfied almost all of the tenants of accepted science. Namely, it was a 'fact' that could be observed by practically everyone who desired to do so. And, so long as we were only discussing large bodies in space, it seemed to be predictable.

All seceding scientists adhered to Newton's theories because they had appeared to hold up. And the parts which did not hold up were, for the most part, ignored or shrugged off. Instead of looking for an alternate explanation, it was easier to look for the 'proverbial' exception to the rule. In the final analysis, that turned out to be most unscientific.

Not to detract from the very important work of talented men like Lorentz, Poincare, Planck, Hertz, etc., etc., but it seems to me absolutely amazing that nobody sought to challenge Newton. Had they done so, how very different science would be today.

My theories are not so much the result of extraordinary intelligence as they are the product of a recalcitrant mind. Of course, it didn't hurt the cause any that I was spared the 'higher education' that my predecessors reveled in. Who knows what damages a college education may have done to me?

I have found one thing that is peculiar to all living beings. No matter how much we learn, there is always going to be more to learn. Take, for instance, the ant. He goes out in search of food. When he discovers it, he calls in his buddies to harvest it. Then, he continues with his quest.

I find that particularly reassuring. How redundant the world would be if we knew everything there was to know. Every quest is like shaking a Christmas gift. We guess at what is inside and, in that precise moment while we are untying the ribbon, it doesn't really matter what is inside the box. We have the ability to imagine anything we want.

Here's hoping that you never stop shaking the boxes. And, if you do stop, let it be because you found whatever it was you were after!

johnhildrethatkins@yahoo.com

www.ingramcontent.com/pod-product-compliance
Lightning Source LLC
Chambersburg PA
CBHW032003170526
45157CB00002B/522